ANTHONY SLY obtained his degree at the University of
Reading and haordshire
schools and coll_____hematics
Department at Q_____ is also a
mathematics exa_____istory of
Computing (Her_____ory Unit
for Computer-based Education). He is also the author of
Modern Mathematics Model Answers and *Modern
Mathematics O-level Passbook* in the Key Facts series.

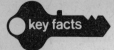
key facts

Multiple Choice

BIOLOGY, J. M. Kelly, BSc

BRITISH ISLES GEOGRAPHY,
D. Bryant, BSc and R. Knowles, MA

CHEMISTRY, C. W. Lapham, MSc

ECONOMICS, J. E. Waszek, BSc(Econ)

ENGLISH LANGUAGE, R. C. Wilson, MA

FRENCH, I. Bryden, BA

GEOGRAPHY, N. E. Law, BA

HISTORY, SOCIAL & ECONOMIC,
S. E. Haworth, BA and M. C. James, BA

HUMAN BIOLOGY,
S. R. Cantle, BSc MMedSci

MODERN MATHEMATICS, A. J. Sly, BA

PHYSICS, B. P. Brindle, BSc

GCE O-level and CSE

Modern Mathematics

A. J. Sly, BA

Published by Charles Letts and Co Ltd
London, Edinburgh and New York

Published 1983 by Charles Letts & Co Ltd.
Diary House, Borough Road, London SE1 1DW

1st edition 1st impression
© Charles Letts & Co Ltd
Made and printed by Charles Letts (Scotland) Ltd
ISBN 0 85097 577 8

Contents

Introduction

You may, or may not, have to answer multiple choice questions as part of your GCE or CSE examinations, but doing so is one of the best ways of finding out whether or not you understand a topic thoroughly. Alongside the correct answer there are three or four other answers intended to distract you.

In compiling the questions for this book, I have followed very closely the order of topics in the *GCE O-level Passbook Modern Mathematics*. The theory and correct method of the solution of any question can be found there. In both books, the questions are modelled on those which appear in the recent GCE and CSE examination papers. When I wrote the Passbook, I anticipated the drawing together of the Modern and Traditional syllabuses which is now taking place. The topics included represent a common core, the best of the old and new, I hope. Consequently, the questions in this book also cover the topics common to most GCE and CSE syllabuses. I have noted carefully the frequency with which topics appear in the multiple choice papers and have attempted to reflect this in the number of questions set on each topic.

All of the questions are intended to be solved without the aid of a calculator, slide rule or tables. Where necessary squares, square roots, logarithms and trigonometrical ratios are given. In many cases the use of any aids would destroy the value of the question. This is in line with the requirements of many examining boards who set a section or whole paper of short answer questions in which no aids are permitted.

Short answer examinations are set to test the candidates' ability to recall facts, to calculate (hence the exclusion of calculators etc,), to solve familiar problems and to apply basic principles to solve unfamiliar problems. The multiple choice examination is such a paper and in it you will find two types of question.

The first, and more common type, is called **simple completion**. A question is set, followed by four or five suggested answers. (For all questions in this book, I have offered five, labelled A, B, C, D and E). Only one answer is correct, but sometimes letter E suggests that none of the offers is correct. You must record the letter of the answer which you think is correct. Here is an example.

1 $\frac{1}{8} + \frac{5}{12}$ is

   A $\frac{6}{20}$ B $\frac{6}{96}$ C $\frac{13}{24}$ D $\frac{17}{96}$ E $\frac{5}{20}$

Alternatively, an incomplete statement is given with five suggested answers. Only one completes this correctly, e.g.,

2 If the length of the side of a square is doubled, the area of the square is

   A halved.   B doubled.   C tripled.   D multiplied by 4.
   E none of them.

The second type of question is called **multiple completion.** Three statements are made, but only one combination of them, lettered A, B, C, D or E, is correct. Here are two examples.

3 In a parallelogram which of the following statements must be true?
   1 The diagonals are equal.
   2 The diagonals bisect each other.
   3 The opposite angles are equal.

   A 1 only   B 2 only   C 1,3   D 1,2,3   E 2,3

4 If $a = (0 \cdot 2)^{-1}$, $b = 25^{\frac{1}{2}}$, $c = 5^0$, which of the following statements is (are) true?

   1 $a = b$ 2 $c > a$ 3 $b < 5c$

   A 1 only   B 2 only   C 3 only   D 2,3   E 1,3

To answer multiple choice questions, do not just look at the question and pick the first answer that appeals to you. This is often no more than a guess which has a $\frac{1}{5}$ chance of success. Read the question and its five suggestions thoroughly. Now be prepared to work out the problem on paper, as you would an ordinary direct question. The answers to the four questions above are now shown with some hints on how to do the working.

1 This is simple completion with only one answer. Work out the sum in the usual way by expressing both fractions in terms of the LCM, in this case 24. Hence, $\frac{1}{8} + \frac{5}{12} = \frac{3}{24} + \frac{10}{24} = \frac{13}{24}$. The correct letter is C. Now scan the other answers again to see if it is possible that the selection is wrong. In A and E the denominators have

been added; a completely wrong approach. In B and D, the denominator of $8 \times 12 = 96$ is possible, so recheck your working to find that C is correct.

2 Again a simple completion answer. It is an incomplete statement, but is best worked out on paper. (Do not guess. There is one very tempting distractor here!) If the side of the square is x, the area of the square is x^2. When the side is doubled it becomes $2x$ and the area becomes $2x \times 2x = 4x^2$. The area of the original has been multiplied by 4. The correct letter to record is D. I think that many candidates, who had not thought carefully, would have chosen B, incorrectly.

3 This is a multiple completion question on the properties of a parallelogram. In any parallelogram the diagonals bisect each other and the opposite angles are equal, so 2 and 3 are true. 'Equal diagonals' is a necessary condition for a rectangle, but not a parallelogram, so 1 is false. 2 and 3 is the correct combination and E is the correct letter to record.

4 This is another multiple completion question. In order to interpret the statements, it is necessary to rewrite the values of a, b and c.

$a = \left(\dfrac{1}{5} \right)^{-1} = 5$, $b = \sqrt{25} = 5$, and $c = 5^0 = 1$.

Now read the statements again, to find that 1 is true, 2 is false, and 3 is false because $b = 5c$. Therefore, the correct combination is 1 only and A is the correct letter to record.

Sometimes it is possible to get the correct answer by working directly and also by working back from the five answers, e.g.,

5 The factors of $x^2 - x - 20$ are

    A $(x + 5)(x - 4)$ B $(x - 5)(x + 4)$ C $(x - 5)(x - 4)$.
    D $(x - 10)(x + 2)$ E $(x + 10)(x - 2)$.

Work directly and put the expression into two brackets, $(x \ ?)(x \ ?)$. The two numbers in the brackets must have a product of -20 and a sum of -1. They are -5 and 4. The factors are $(x - 5)(x + 4)$. Alternatively, multiply out each pair of brackets in A, B, C, D and E to find that
$(x - 5)(x + 4) = x^2 - 5x + 4x - 20 = x^2 - x - 20$,
and B is confirmed as the correct answer.

Chapter 1
Sets

1 $A = \{m, a, t, h, s\}$ $B = \{e, x, a, m, s\}$ $C = \{m, a, s, t, e, r\}$,
   $n(A \cup B \cup C)$ is

   A 3   B 8   C 5   D 16   E none of them.

2 P is the set of prime numbers between 70 and 90; $n(P)$ is

   A 6   B 7   C 5   D 8   E 4

3 A set S contains 10 elements and a set T contains 14 elements.
   The least possible number of elements in $S \cap T$ is

   A 0   B 4   C 10   D 14   E 24

4 Given that $n(P) = 7$ and $n(Q) = 11$, the least possible value of
   $n(P \cup Q)$ is

   A 7   B 11   C 18   D 4   E 77

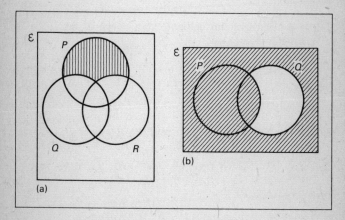

Figure 1

5 In figure 1(a), the shaded area is

   A P B $P \cap Q \cap R'$ C $P \cap Q' \cap R'$
   D $R' \cup (P \cup Q)$ E $P \cup Q$

9

6 In figure 1(b), the shaded area is

A $P \cup Q'$ B $P \cap Q'$ C $P' \cup Q'$ D $P' \cap Q$
E $(P \cup Q)'$

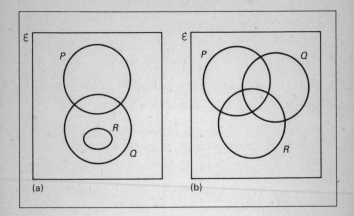

Figure 2

7 In the Venn diagram in figure 2(a), which of the following statements is (are) true?

1 $R \subset Q$
2 $R \cup Q = R$
3 $R \subset P'$

A 1 only B 1,2 C 1,2,3 D 1,3 E 2,3

8 In figure 2(b), P,Q and R are sets. Which one of the following is equivalent to $P \cup (Q \cap R)$?

A $P \cap (Q \cup R)$ B $(P \cap Q) \cup (P \cap R)$
C $(P \cup Q) \cap (P \cup R)$ D $P' \cap (Q \cup R)$
E $P \cap (Q' \cup R)$

9 If $P = \{5,7,9,11\}$, $Q = \{4,7,11\}$, $R = \{1,2,4,7\}$, then

1 $(P \cap Q) \subset R$
2 $(P \cap R) \subset Q$
3 $(Q \cap R) \subset P$

A 1,3 B 1,2 C 2,3 D 1,2,3 E 2 only

10 $X = \{x : -5 \leqslant x < 3\}$, $Y = \{x : -3 < x \leqslant 5\}$. $X \cap Y$ is

A $\{x : -3 < x \leqslant 5\}$. B $\{x : -5 \leqslant x \leqslant 5\}$. C $\{x : -3 \leqslant x \leqslant 3\}$.
D $\{x : -5 \leqslant x < 3\}$. E $\{x : -3 < x < 3\}$.

11 $S = \{\text{unmarried people}\}$, $M = \{\text{men}\}$,
$R = \{\text{right-handed people}\}$. Which of the following represents
the set of married men who are left-handed?

A $M \cap S' \cap R'$ B $M \cup S' \cup R'$ C $M \cap (S' \cup R')$
D $S \cap M' \cap R'$ E $S' \cap M$

12 $P = \{\text{parallelograms}\}$, $R = \{\text{rhombuses}\}$, $S = \{\text{squares}\}$. Which
of the following is (are) **false?**

1 $R \subset S$
2 $R \subset P$
3 $P \subset S$

A 1,3 B 1 only C 2,3 D 1,2,3 E 2 only

13 $\varepsilon = \{52 \text{ cards of a pack of playing cards}\}$
$A = \{\text{the four aces}\}$
$R = \{\text{all the red cards}\}$
$B = \{\text{all the black cards}\}$. Then $A \cap R$ is

A $A' \cap R'$ B $A' \cap R$ C $A \cap B'$ D $A' \cap B$ E $A \cap B$

14 $\varepsilon = \{\text{natural numbers}\}$ $P = \{\text{multiples of 3}\}$
$Q = \{\text{multiples of 4}\}$ $R = \{\text{multiples of 10}\}$
Then $P \cap Q \cap R'$ represents

A $\{\text{multiples of 40}\}$. B $\{\text{multiples of 120}\}$.
C $\{\text{multiples of 3 which are not multiples of 10}\}$.
D $\{\text{multiples of 12 which are not multiples of 10}\}$.
E $\{\text{multiples of 6 which are not multiples of 5}\}$.

15 In a group of 32 people, 22 like Indian food, 12 like Chinese
food and 3 people like neither. How many people like Chinese
food but do not like Indian food?

A 5 B 17 C 7 D 20 E 10

16 In a school of 1000 pupils, 700 study Physics, 600 study
Chemistry and 400 study both. The number who studies
neither is

A 300 B 250 C 200 D 150 E 100

17 $Q' \cup (P \cap Q) =$

   A P B ε C $Q' \cap P$ D $P \cup Q$ E $Q' \cup P$

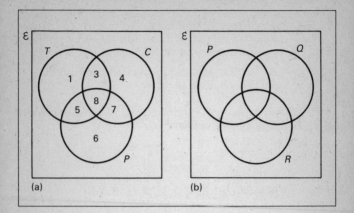

Figure 3

18 In figure 3(a), the numbers in the spaces represent the number of elements in each of the sets. If $n(\varepsilon) = 40$, which of the following statements is (are) true?

  1 $n(C \cap T) = 11$
  2 $n(C \cup P \cup T)' = 6$
  3 $n(C) = 4$

  A 1 only B 3 only C 2 and 3 D 1 and 2
  E none of them.

19 In figure 3(b), $n(\varepsilon) = 30$, $n(P) = 13$, $n(Q) = 17$, $n(R) = 25$, $n(P \cap Q) = 9$, $n(P \cap R) = 11$, $n(Q \cap R) = 12$ and $n(P \cup Q \cup R)' = 0$.

  $n(P \cap Q \cap R)$ is

  A 25 B 2 C 7 D 0 E none of them.

Chapter 2
Basic Algebra

1 Given that $a = -5$, $b = 2$, the value of $a^2 - ab$ is

   A -35 B -15 C 0   D 15   E 35

2 If $x = 10$, the value of $\frac{3}{5}(x-5) - \frac{1}{2}(4-x)$ is

   A 0   B 6   C 12   D 10   E 9

3 If $x = 3^a$ and $y = 3^b$, then xy is equal to

   A 3^{a+b} B 3^{ab} C $a - b$ D $a + b$ E 9^{a+b}

4 In the formula $P = \dfrac{5a^3}{b}$, a and b are both doubled. The value of P is

   A multiplied by 8.   B multiplied by 16.
   C multiplied by 4.   D doubled.   E unchanged.

5 If p is doubled and q is halved, then the value of $8pq^2$ is

   A unchanged.   B halved.   C divided by 4.   D doubled.
   E multiplied by 4.

6 The product of $3a$, a, $-2a^2$, is

   A $-6a^4$ B $-5a^4$ C a^2 D $4a - 2a^2$ E $-6a^2$

7 $4(a+3b) - 3(a-4c) - 5(c-2b)$ is

   A $a + 2b + 7c$ B $a + 2b + 17c$ C $a + 22b - 17c$
   D $a + 22b + 7c$ E $a + 22b - 7c$

8 $(a-b)^2$ is

   A $a^2 - b^2$ B $a^2 + b^2$ C $a^2 + 2ab + b^2$ D $a^2 - 2ab - b^2$
   E $a^2 - 2ab + b^2$

9 $(x-2)^2 + x - 2$ is

   A $(x-2)(x-3)$ B $(x-2)(x-1)$ C $(x-2)^2$
   D $(x-2)(x+1)$ E $2(x-2)$

10 By how much does $(a+2)^2$ exceed $(a-2)^2$?

   A $4a$ B $-8a$ C $8a$ D $-4a$ E 4

11 $(a-b)^2 - (a^2 - b^2)$ is

   A $2a^2 - 2ab$ B $2b^2 - 2ab$ C 0
   D $-2b^2$ E $2b^2$

12 If $4x^2 - 12x + k$ is a perfect square, k is

   A -9 B $+9$ C -16 D $+16$ E 1

13 The factors of $x^2 - 5x - 6$ are

   A $(x+6)(x-1)$ B $(x-6)(x+1)$ C $(x-3)(x-2)$
   D $(x+3)(x-2)$ E $(x-3)(x+2)$

14 The factors of $3x^2 - 2x - 8$ are

   A $(3x-2)(x+4)$ B $(3x+4)(x-2)$ C $(3x+2)(x-4)$
   D $(3x-4)(x+2)$ E $(3x-8)(x+1)$

15 If $(x+2)$ is a factor of $2x^2 - x + k$, k is

   A -12 B $+6$ C -10 D -6 E $+10$

16 The factors of $ab + xy - ay - xb$ are

   A $(a+x)(b-y)$ B $(a-x)(b-y)$ C $(a-x)(y+b)$
   D $(a+x)(y-b)$ E $(x-a)(b-y)$

17 Which of the following is a common factor of $x^2 - 5x + 4$ and
   $x^2 - x - 12$.

   A $(x+4)$ B $(x-1)$ C $(x-4)$
   D $(x+1)$ E $(x-2)$

18 The factors of $(a+b)^2 - (a^2 - b^2)$ are

   A $2a(a+b)$ B $2b(a+b)$ C $(a+b)(a-b)$
   D $(a+b)^2(a-b)$ E $2b^2$

19 $6x^2 - 11x - 10$ is exactly divisible by

   1 $(3x+2)$ 2 $(2x+5)$ 3 $(3x-5)$

   A 1 only B 2 only C 3 only
   D 2,3 E 1,2

20 $\dfrac{b}{a} - \dfrac{a}{b}$ is

   A $\dfrac{b-a}{ab}$ B $\dfrac{a-b}{ab}$ C $\dfrac{a^2-b^2}{ab}$ D $\dfrac{b^2-a^2}{ab}$ E -1

21 $\dfrac{a}{6} + \left(\dfrac{b}{3}\right)^2$ when expressed as a single fraction is

   A $\dfrac{a+b^2}{6}$ B $\dfrac{3a+2b^2}{18}$ C $\dfrac{ab^2}{18}$ D $\dfrac{9a+b^2}{54}$

   E $\dfrac{6x-1}{12}$

22 $\dfrac{\dfrac{1}{a}+b}{a+\dfrac{1}{b}}$ can be simplified to

   A 1   B 2   C $a^2+\dfrac{1}{b^2}$ D $b^2+\dfrac{1}{a^2}$ E $\dfrac{b}{a}$

23 $\dfrac{2a}{3b} + \dfrac{3a}{5b}$ is

   A $\dfrac{19a}{15b}$ B $\dfrac{19a}{15b^2}$ C $\dfrac{5a}{8b}$ D $\dfrac{5a}{8b^2}$ E $\dfrac{5a^2}{8b^2}$

24 $\dfrac{5x}{4} - \dfrac{x-1}{3}$ simplifies to

   A $\dfrac{5x+1}{12}$ B $\dfrac{5x+4}{12}$ C $\dfrac{11x+1}{12}$ D $\dfrac{11x+4}{12}$ E $\dfrac{11x-4}{12}$

25 $\dfrac{2x^4 + 10x^2}{2x}$ when simplified, becomes

   A x^3+5x B x^3+10x C $2x^4+5x$ D $2x^4+5x^2$
   E x^3+8x

26 When simplified, $\dfrac{4x^2-9}{2x-3}$ becomes

   A $2x-3$ B $2x^2-3$ C $2x-6$ D $2x+3$ E none of them.

27 $\dfrac{3x-2}{2} - \dfrac{2x-3}{4}$ simplifies to

   A $\dfrac{4x-1}{4}$ B $\dfrac{4x-7}{4}$ C $\dfrac{4x-5}{4}$ D $\dfrac{4x+1}{4}$ E $x-1$

28 If $x - 2(3x - 5) = 20$, the value of x is

A $-12\frac{1}{2}$ B -2 C -6 D $-7\frac{1}{2}$ E -3

29 $\frac{1}{2} - \frac{x+3}{3} = 1$, has as its solution $x =$

A $-4\frac{1}{2}$ B -2 C $1\frac{1}{2}$ D 4   E -4

30 If $\frac{4}{x} = 3 + \frac{3}{x}$, then x is

A $2\frac{1}{3}$ B 3   C $\frac{1}{3}$ D $\frac{3}{7}$ E $-\frac{3}{7}$

31 If $6x + 5y = 25$
       $3x - 5y = 5$, then

A $2x = 5$ B $3x = 20$ C $3x = 30$ D $9x = 20$ E $9x = 30$

32 If $2x - 3y = 2$
       $3x - 2y = -2$, the value of y is

A $-\frac{10}{13}$ B -2 C $-\frac{2}{13}$ D $+2$ E $-\frac{4}{5}$

33 If $3x - 2y = 5$
       $2x - y = -2$, then $x - y$ is

A $+7$ B -7 C $+3$ D -3 E $-2\frac{1}{2}$

34 Given that x is a positive integer and $x^2 - 6x - 16 = 0$, then the value of x is

A 8   B 2   C 4   D 16   E 1

35 The roots of an equation are $-\frac{3}{4}$ and $\frac{2}{3}$. The equation is

A $(3x - 4)(2x - 5) = 0$ B $(4x - 3)(5x - 2) = 0$
C $(3x + 4)(2x + 5) = 0$ D $(4x + 3)(5x - 2) = 0$
E $(3x + 4)(2x - 5) = 0$

36 The quadratic equation $x^2 + px + q = 0$ has solutions $x = 1$ and $x = -3$. The value of p is

A 2   B -3 C -2 D 4   E -4

37 The equation $(x - 2)(x - 2) + (x - 2)(x - 1) = 0$
   has two solutions. The smaller value of x is

A 2   B $1\frac{1}{2}$ C -2 D 1   E $\frac{1}{2}$

16

38 The solution set of $2x^2 - 5x = 0$ is

   A $\{2\frac{1}{2}\}$ B $\{0, 2\frac{1}{2}\}$ C $\{0, \frac{2}{5}\}$ D $\{0, 2\frac{1}{2}, \frac{2}{5}\}$ E $\{0\}$

39 If $x^2 - 3 = 0$ and $x = 1 - t$, then t satisfies

   A $t^2 - 2t + 4 = 0$ B $-t^2 + 2t + 4 = 0$ C $-t^2 - 2t - 2 = 0$
   D $t^2 - 2t - 2 = 0$ E $t^2 + 2t - 2 = 0$

40 $(3x - 2)(x + 2) = 3x^2 + 4x - 4$ is a statement that is true for

   A no value of x. B one value of x. C two values of x.
   D three values of x. E all values of x.

41 $(x + 5)^2 = x^2 + 25$ is true for

   A no value of x. B one value of x. C two values of x.
   D three values of x. E all values of x.

42 A car travels a distance of s km in 20 minutes. It travels at the same average speed for h hours. The total distance it travels, in kilometres, is

   A $\dfrac{sh}{3}$. B $3sh$. C $\dfrac{s}{20h}$. D $20sh$. E $\dfrac{5h}{s}$.

43 A car hire firm hires its cars for £P per day plus d pence per mile. The total cost, in pounds, of hiring a car for one day to make a journey of m miles is

   A $P + 100md$. B $P + \dfrac{md}{100}$. C $100P + md$. D $\dfrac{P + md}{100}$.
   E $P + md$.

44 x articles cost $(x + 3)$ pence each and another $2x$ articles cost $4x$ pence each. The average price of all the articles is

   A $\dfrac{5x + 3}{3x}$. B $6x$. C $\frac{1}{2}(3x + 3)$. D $3x + 1$. E $9x^2 + 1$.

45 If $v = u + at$, then t is

   A $\dfrac{v}{u + a}$ B $v - u + a$ C $\dfrac{v - u}{a}$ D $\dfrac{v}{u} - a$ E $\dfrac{v + u}{a}$

46 If $p = q - \dfrac{r}{s}$, then s is

   A $\dfrac{r}{q - p}$ B $\dfrac{q - p}{r}$ C $\dfrac{r}{p - q}$ D $\dfrac{p - q}{r}$ E $\dfrac{r}{p + q}$

47 If $E = mc^2$, then $c =$

A $\pm\sqrt{E-m}$ B $\left(\dfrac{E}{m}\right)^2$ C $\pm\dfrac{\sqrt{E}}{m}$ D $\pm\sqrt{\dfrac{E}{m}}$ E $\pm\sqrt{Em}$

48 If $\sqrt{\left(\dfrac{x-3a}{4}\right)} = 2a$, then x is

A $11a$ B $5a$ C $3a + 16a^2$ D $19a$ E $3a + 8a^2$

49 If $a = \dfrac{b+2}{b-2}$, then b is

A -1 B -2 C $\dfrac{4}{a-1}$ D $\dfrac{a+2}{a-1}$ E $\dfrac{2a+2}{a-1}$

50 If $\dfrac{1}{f} = \dfrac{1}{u} + \dfrac{1}{v}$, then $f =$

A $u + v$ B $\dfrac{uv}{2}$ C $\dfrac{uv}{u+v}$ D $v - u$ E $\dfrac{u+v}{u-v}$

51 If $\dfrac{p+2q}{p} = \dfrac{5}{2}$, then the value of $\dfrac{p}{q}$ is

A $\dfrac{5}{4}$ B $\dfrac{4}{7}$ C $\dfrac{3}{4}$ D $\dfrac{4}{3}$ E none of them.

52 If $3b = 5a$, then $\dfrac{a+b}{a-b}$ is

A 4 B -3 C -4 D $+3$ E none of them.

53 If $\dfrac{2}{3} = \dfrac{a}{15} = \dfrac{8}{b}$, then $a : b$ is

A $6 : 5$ B $5 : 6$ C $15 : 8$ D $8 : 15$ E $4 : 5$

54 Given that $2x^2 - 9x - 3 = 0$, then x is

A $\dfrac{9 \pm \sqrt{57}}{4}$ B $\dfrac{-9 \pm \sqrt{105}}{4}$ C $\dfrac{9 \pm \sqrt{105}}{4}$

D $\dfrac{-9 \pm \sqrt{57}}{4}$ E none of them.

Chapter 3
Number Systems

1 The difference between the greatest and the least prime numbers between 70 and 88 is

  A 8 B 10 C 12 D 14 E 16

2 p and q are positive integers. $p - q$ **must** be a number which is

  A natural. B an integer. C irrational. D positive.
E prime.

3 p is a natural number. Which of the following **must** be an odd number?

  A $\frac{1}{2}p$ B $2p-1$ C $(p+1)^2$ D p^2

  E $\frac{p}{2}(p+1)$

4 No prime number can have as its last digit a

  A 2 B 3 C 5 D 4 E 7

5 If n is an integer and $\frac{n+1}{2}$ is also an integer, then n **must** be an integer that is

  A negative. B positive. C even. D odd.
E irrational.

6 If $\varepsilon = \{3\frac{1}{7}, 3.14, \pi, -3\frac{1}{7}\}$ and $R = \{\text{rational numbers}\}$, then the set R is

  A $\{3\frac{1}{7}, 3\cdot14, -3\frac{1}{7}\}$. B $\{3\frac{1}{7}, -3\frac{1}{7}\}$. C $\{3\frac{1}{7}, 3\cdot14\}$.
D $\{3\frac{1}{7}, 3\cdot14, \pi\}$. E ε.

7 $2^5 + 2^3 + 2$ can be written as the binary number

  A 11 000 B 10 101 C 10 110 D 101 010 E 10 111

8 The denary number 40, written as a base 4 number, is

  A 11 000 B 2100 C 220 D 230 E none of them.

9 The denary number 37 is the binary number

   A 10 101   B 100 101   C 101 001   D 11 001   E 10 111

10 32_8 written in base 2 is

   A 11 010   B 110 111   C 10 110   D 111 100   E 11 001

11 If $122_5 - 34_5 = x_{10}$, the value of x is

   A 3   B 88   C 22   D 7   E 18

12 If $224_n + 144_n = 412_n$, then n is

   A 4   B 5   C 6   D 7   E 8

13 The smallest of 100_{10}, 110_7, 1100_3, $100\ 101_2$, 1010_4, is

   A 100_{10} B 110_7 C 1100_3 D $100\ 101_2$ E 1010_4

14 $a = 1011_2$ $b = 1011_3$ $c = 13_8$ $d = 111_4$.
Which of the following is (are) true?

  1 $a = c$
  2 $d > b$
  3 $\frac{1}{2}(b + c) = d$

   A 1 only   B 3 only   C 2,3   D 1,2   E 1,3

15 Which of the following is (are) odd?

  1 112_3
  2 122_3
  3 211_3

   A 3 only   B 1,3   C 2 only   D 2,3   E 1,2

16 The area of a square is $11\ 001_2$ cm^2. The perimeter, in cm, is

   A 1010_2 B $101\ 000_2$ C $110\ 010_2$ D $10\ 100_2$
   E none of them.

17 The mean of three base 4 numbers is 13_4. Two of the numbers are 3_4 and 20_4. The third number is

   A 13_4 B 22_4 C 12_4 D 111_4 E 11_4

18 If $5 + 4 = x$ (mod 6), then x is

   A 1   B 2   C 3   D 4   E 5

19 If $5 \times 4 = x$ (mod 7), then x is

   A 0 B 2 C 4 D 5 E 6

20 If $3x = 3$ (mod 6), then x is

   A 1 B 3 C 1 and 3 D 1 and 5 E 1,3 and 5

21 If $2 - 1 - 3 = x$ (mod 5), then x is

   A -2 B 4 C 1 D 2 E 3

22 Given that $a * b$ means "add 2 to a and multiply the result by b", the value of $(3 * 2) * 4$ is

   A 48 B 24 C 40 D 56 E 35

23 The operation $*$ is defined by $a * b = a^2 + 2b$. If $4 * x = 26$, the value of x is

   A 21 B 8 C 22 D 10 E 5

24 The table defines the operation $*$ on the set $\{p,q,r,s,t\}$.

*	p	q	r	s	t
p	r	q	p	s	t
q	q	p	q	p	q
r	p	q	s	r	s
s	t	p	r	s	q
t	r	s	p	q	t

The value of $(q * r) * p$ is

A p B q C r D s E t

25

*	a	b	c	d
a	d	a	b	c
b	a	b	c	d
c	c	c	d	a
d	d	d	a	b

In this operation table, the identity element is

A a B b C c D d E none of them.

26

*	a	b	c	d
a	b	a	a	c
b	a	a	d	c
c	a	d	b	a
d	c	c	a	d

The table defines the operation * on the set $A = \{a, b, c, d\}$.
Which of the following statements is (are) true?

1 c is the identity element of A.
2 A is closed under the operation *.
3 The operation is commutative.

A 1,2   B 1,3   C 2,3   D 3 only   E 2 only.

27 The operation * on a set is defined as $a * b = ab + 1$.
Which of the following statements is (are) true?

1 The operation is commutative.
2 The operation is associative.
3 $x * (y + z) = x * y + x * z$.

A 1 only   B 2 only   C 3 only   D 2,3   E 1,2

28

*	p	q	r	s
p	s	p	s	p
q	p	q	r	s
r	q	r	q	r
s	r	s	p	q

The table defines the operation * on a set $A = \{p, q, r, s\}$.
Which of the following statements is (are) true?

1 The operation * is associative.
2 q is the identity element of A.
3 The inverse of s is s.

A 1,2,3   B 1,2   C 2,3   D 1,3   E 2 only.

Chapter 4
Functions (1)

1 Under the mapping $f : x \to x^3$, the domain is $0 \leqslant x \leqslant 8$.

The range is

A $-2 \leqslant f(x) \leqslant 2$. B $0 \leqslant f(x) \leqslant 2$. C $-8 \leqslant f(x) \leqslant 8$.
D $0 \leqslant f(x) \leqslant 512$. E $-512 \leqslant f(x) \leqslant 512$.

2 Under the mapping $f : x \to x^2$, the range is the set of values of y given by $0 \leqslant y \leqslant 4$. The domain is the set of values of x given by

A $0 \leqslant x \leqslant 2$. B $-2 \leqslant x \leqslant 2$. C $0 \leqslant x \leqslant 16$.
D $-16 \leqslant x \leqslant 16$. E $-16 \leqslant x \leqslant 0$.

3 Under the mapping $f : x \to 3^x$, the domain is $-2 \leqslant x \leqslant 2$.

The range is the set of values of y given by

A $-6 \leqslant y \leqslant 6$. B $-\frac{1}{6} \leqslant y \leqslant 6$. C $\frac{1}{6} \leqslant y \leqslant 6$.
D $\frac{1}{9} \leqslant y \leqslant 9$. E $0 \leqslant y \leqslant 9$.

4 Given that $f(x) = x^3 - 5x^2$, $f(-2)$ is

A -12 B -28 C $+28$ D $+12$ E -26

5 If $f(x) = \dfrac{x^2 - 1}{3x + 4}$, then $f(-2) =$

A $\frac{3}{2}$ B $\frac{1}{2}$ C $-\frac{3}{2}$ D $-\frac{1}{2}$ E $\frac{3}{10}$

6 Given that $f(x) = 2x^3 + 6$, the average of $f(3)$ and $f(0)$ is

A 30 B 33 C 15 D 12 E 114

7 If $f(x) = 2x^3 - x^2 - k$ and $f(3) = 41$, then k is

A -4 B 86 C 4 D -86 E -32

8 A mapping is defined as $f : x \to (x-5)^2$. The image of (-2) under the mapping is

A 49 B 9 C 29 D -9 E -49

9

x	-1	0	1
$f(x)$	2	3	4

Which of the following functions is (are) satisfied by the values in the table?

1 $f: x \rightarrow x + 3$
2 $f: x \rightarrow 2^x + 3$
3 $f: x \rightarrow x^3 + 3$

A 1 only   B 2 only   C 3 only   D 1,3   E 2,3

10 If $f: x \rightarrow x^2$ and $g: x \rightarrow \dfrac{2x+1}{3}$, then $fg(-5) =$

A -9 B 6   C $\dfrac{121}{9}$ D 17   E 9

11 If $f: x \rightarrow \dfrac{1}{x}$ and $g: x \rightarrow (x-1)^2$, then $gf(3)$ is

A $-\frac{4}{9}$ B $\frac{4}{9}$ C $\frac{4}{3}$ D $\frac{3}{4}$ E $\frac{1}{4}$

12 If $f: x \rightarrow x + 3$ and $g: x \rightarrow \dfrac{1}{x}$, then $fg: x \rightarrow$

A $\dfrac{1}{x} + 3$ B $\dfrac{x}{x+3}$ C $x + \dfrac{3}{x}$ D $\dfrac{1}{x+3}$ E $1 + \dfrac{3}{x}$

13 x is real and non zero, f and g are functions of x. Which of the following pairs satisfy the condition that $fg = gf$?

1 $f: x \rightarrow 3x, \ g: x \rightarrow x^2$
2 $f: x \rightarrow x + 3, \ g: x \rightarrow x - 2$
3 $f: x \rightarrow \dfrac{1}{x}, \ g: x \rightarrow \dfrac{x}{2}$

A 1 only   B 2 only   C 3 only   D 1,2   E 1,2,3

14 The inverse of the function $x \rightarrow 4x - 3$ is

A $x \rightarrow 4x + 3$ B $x \rightarrow 4(x+3)$ C $x \rightarrow \frac{1}{4}x + 3$
D $x \rightarrow \frac{1}{4}(x+3)$ E none of them.

15 Given that $f : x \rightarrow 5(3x-2)$, then $f^{-1}(x)$ is

   A $3(5x-2)$ B $3\left(\dfrac{x}{5}+2\right)$ C $\dfrac{1}{3}\left(\dfrac{x}{5}+2\right)$

   D $\dfrac{1}{3}\left(\dfrac{x}{5}-2\right)$ E none of them.

16 If $f : x \rightarrow \dfrac{4}{x}+2$, then $f^{-1}(3)$ is

   A $-1\frac{3}{4}$ B $-2\frac{3}{4}$ C $1\frac{1}{4}$ D $-\frac{1}{4}$ E 4

17 If $f : x \rightarrow 2x-1$ and $g : x \rightarrow 4x-3$, which of the following statements is (are) true?

   1 $ff(x) = g(x)$ 2 $ff^{-1}(x) = g^{-1}(x)$
   3 $f^{-1}(1) = g^{-1}(1)$

   A 1,3 B 2,3 C 1 only D 2 only E 3 only

18 A is the point $(-4,5)$, B is $(2,2)$ and C is $(4,3)$. The ratio $AB : BC$ is

   A $1:3$ B $2:3$ C $3:2$ D $3:1$ E $1:6$

19 A is the point $(-2,3)$ and B is $(3,-1)$. The length of AB is

   A 3 units. B $\sqrt{41}$ units. C 9 units. D $\sqrt{17}$ units.
   E $\sqrt{20}$ units.

20 A is the point $(3,-1)$ and B is $(-2,4)$. The gradient of AB is

   A -1 B 1 C $-\frac{1}{3}$ D $-\frac{5}{3}$ E $\frac{1}{5}$

21 The co-ordinates of A and B are $(1,3)$ and $(5,k)$ respectively. The gradient of AB is $\frac{3}{2}$. The value of k is

   A 3 B 4 C 6 D 9 E 12

22 Which of the following is (are) straight lines?

   1 $y=3x^2+2$ 2 $y=\dfrac{3}{x}+2$ 3 $y=3x+2$

   A 1 only B 2 only C 3 only D 1,2 E 2,3

23 The straight line $y=4x-5$ passes through the point

   A $(5,0)$ B $(0,5)$ C $(-5,0)$ D $(0,-5)$ E $(5,-5)$.

24 The gradient of the straight line $y = \dfrac{4-2x}{3}$ is

A -2 B 2 C $-\frac{2}{3}$ D $-\frac{3}{2}$ E $\frac{4}{3}$

25 In figure 4, which diagram represents the line $3x + 4y = 2$?

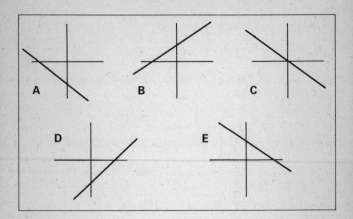

Figure 4

26 The equation of the straight line with gradient -3, which passes through the point $(-3,4)$ is

A $y = 3x + 23$ B $y = 23 - 3x$ C $y = -5 - 3x$ D $y = 3x + 5$
E $3y - x = 15$

27 The straight line $4x + y = 2$ cuts the x-axis at P and the y-axis at Q. If O is the origin, the area of the triangle POQ, in square units, is

A $\frac{1}{2}$ B 1 C $1\frac{1}{2}$ D 2 E 4

28

x	0	P	10
y	0	3	8

The table shows points on a straight line. The value of P is

A $3\frac{1}{3}$ B $3\frac{3}{4}$ C 5 D 3 E $1\frac{1}{4}$

29 The equation of the straight line which passes through the points $(-1,2)$ and $(3,1)$ is

A $4x+y=7$ B $4y+x=7$ C $4y-x+7=0$
D $4x-y-7=0$ E $3x+y=2$

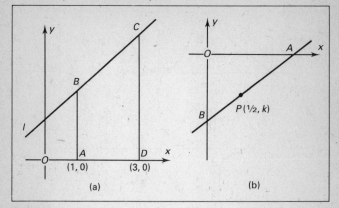

Figure 5

30 If $A=\{(x,y) : 2x+3y=2\}$ and $B=\{(x,y) : 3x-2y=16\}$, then $A \cap B$ is

A $\{(4,2)\}$ B $\left\{\left(\dfrac{44}{5}, \dfrac{-26}{5}\right)\right\}$ C $\{(2,-2)\}$ D $\{(4,-2)\}$

E $\left\{\left(\dfrac{52}{5}, \dfrac{-38}{5}\right)\right\}$

31 In figure 5(a), the straight line is $y=2x+3$. The co-ordinates of A and D are $(1,0)$ and $(3,0)$ respectively. The area of $ABCD$, in square units, is

A 28 B 21 C 11 D 7 E 14

32 In figure 5(b), the straight line is $3y-2x+4=0$. The point P has co-ordinates $(\frac{1}{2},k)$. Which of the following statements is (are) true?

1 The co-ordinates of B are $(0,-\frac{2}{3})$.

2 The co-ordinates of A are $(2,0)$.

3 The distance of P from the origin is $\sqrt{\dfrac{5}{4}}$.

A 1 only B 2 only C 1,3 D 2,3 E none of them.

Chapter 5
Functions (2)

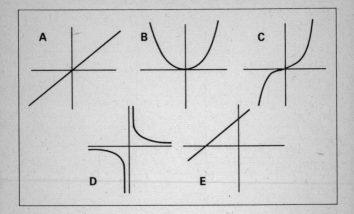

Figure 6

1 In figure 6, which of the graphs represents the function $y = kx^2$, where k is a positive constant?

2

x	4	8
y	7	2

Which of the following relations between x and y is (are) possible from the above table?

1 $y = ax + b$ 2 $y = kx^2$ 3 $y = \dfrac{P}{x}$

A 1only B 2 only C 3 only D 1,2 E 2,3

3 The curve $y = x^2 - 7x - 3$ cuts the y-axis at the point

A $(0,1)$ B $(0,-7)$ C $(0,7)$ D $(0,-3)$ E $(0,3)$

4 The curve whose equation is $y = 6 - \dfrac{5}{x}$ passes through the point $(a,2)$. The value of a is

A $1\frac{1}{4}$ B $\frac{4}{5}$ C $-\frac{4}{5}$ D $1\frac{1}{3}$ E $-1\frac{3}{5}$

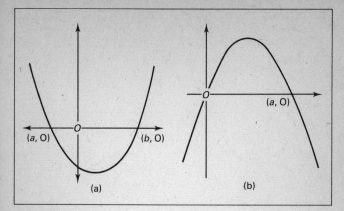

Figure 7

5 In figure 7(a), the graph of $y = x^2 - 3x - 4$ is drawn. The values of a and b are respectively

   A -4 and 3   B -2 and 2   C -1 and 4   D 1 and -4
   E -3 and 4

6 Figure 7(b) represents the graph of $y = 5x - 2x^2$. The value of a is

   A $\frac{2}{5}$ B $\frac{5}{2}$ C 3   D 4   E none of them.

7 The graphs of $x \to 2x + 1$ and $x \to 2x^2 + x$ are drawn on the same axes. The number of points of intersection is

   A 0   B 1   C 2   D 3   E 4

8 In figure 8(a), the parabola which cuts the x-axis at the points A and B is the function

   A $f : x \to (x - 1)(2x + 5)$.   B $f : x \to (x + 1)(2x - 5)$.
   C $f : x \to (x - 1)(5x - 2)$.   D $f : x \to (x + 1)(5x + 2)$.
   E $f : x \to (2x + 5)(x + 1)$.

9 In figure 8(b), l is a straight line and the curve is of the form $y = ax^2 + bx + c$. They intersect at A and B. The x co-ordinates of A and B are given by the equation

   A $x^2 - 3x - 10 = 0$.   B $x^2 - 3x - 3 = 0$.   C $5x - 2x^2 = 0$.
   D $x^2 - 7x + 10 = 0$.   E none of them.

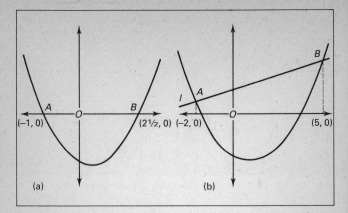

Figure 8

10 y is directly proportional to x. When $x = 2$, $y = 10$. When $y = 7$
the value of x is

A 2   B 1·4   C 35   D 1·2   E 19

11 y varies as the square of x, and $y = 3$ when $x = 2$. When $x = 6$
the value of y is

A 48   B $4\frac{1}{2}$ C 27   D 9   E 54

12 If p and q are two variables such that p varies inversely as the
square root of q, then q varies

A directly as the square of p.
B inversely as p.
C directly as the square root of p.
D inversely as the square root of p.
E inversely as the square of p.

13 L varies jointly as the square of M and inversely as the cube of
R.
If k is the constant of variation, then $L = $

A $\dfrac{k\sqrt{M}}{R^3}$ B $\dfrac{kR^3}{M^2}$ C $\dfrac{kM^2}{R^3}$ D kM^2R^3

E none of them.

14 A car travels 100 km in $2\frac{1}{2}$hr. Its average speed, in km per hour, is

   A 20   B 30   C 40   D 50   E 60

15 A train travels 240 km at an average speed of 75 km per hour. The time taken is

   A 4 hours.   B 3 hours 20 minutes.   C 4 hours 20 minutes.
   D 3 hours 2 minutes.   E 3 hours 12 minutes.

16 When a car is travelling at 36 km per hour its speed, in cm per second, is

   A 100   B 1000   C 10 000   D 200   E 2000

17 When a car is travelling at an average speed of 25 metres per second its speed, in km per hour, is

   A 15   B 60   C 90   D 45   E 75

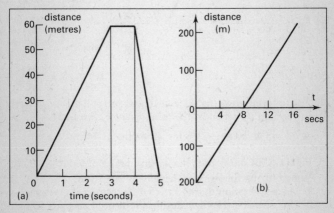

Figure 9

18 Figure 9(a) shows the distance-time graph of a moving object. Which of the following statements is (are) true?
   1 The object is at rest during the fourth second.
   2 The object slows down during the fifth second.
   3 The average speed during the first three seconds is 20 metres per second.

   A 1 only   B 2 only   C 1,2   D 1,3   E 1,2,3

19 The distance-time graph in figure 9(b), for a moving object represents

A the object coming to rest after 8 seconds.
B the object accelerating.
C the object travelling 200 metres in 16 seconds.
D the average speed of the object is 25 metres per second.
E none of them.

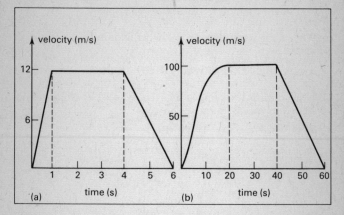

Figure 10

20 Figure 10(a) shows the velocity-time graph of a moving object. In the first six seconds, the total distance travelled is

A 72 m   B 60 m   C 56 m   D 48 m   E 54 m

21 Figure 10(b) shows the velocity-time graph of a moving body. Which of the following statements is (are) true?
1 The acceleration during the first 20 seconds is constant.
2 The distance travelled during the last 20 seconds is 1000 metres.
3 The maximum speed is 18 km per hour.

A 1 only   B 2 only   C 3 only   D 1,3   E 2,3

Chapter 6
Inequalities

1 If $3 \leqslant 2x - 5 < 13$, then

  A $8 \leqslant x < 18$. B $4 \leqslant x < 9$. C $4 \leqslant x \leqslant 9$. D $4 < x \leqslant 9$.
  E none of them.

2 Given that $5 - 4x > 13$, then

  A $x < 4\frac{1}{2}$. B $x > -2$. C $x > 2$. D $x < -2$. E $x < 2$.

3 The inequality $4x + 20 < 4(x + 3)$ is true for

  A $x < -1$. B $x < 4$. C $x > 4$. D no values of x.
  E any value of x.

4 The inequality $3x + 8 < 3(x + 4)$ is true for

  A $x < -2$. B $x > 2$. C $x < 2$. D no value of x.
  E any value of x.

5 If x is an integer and $4 - 5x < 2 - 7x$, the greatest value of x to satisfy the inequation is

  A -1 B -2 C 0 D 1 E 2

6 If $p - 2(2p + 1) > 7$, then

  A $p < -\frac{5}{3}$. B $p > \frac{5}{3}$. C $p > 3$. D $p < -3$.
  E none of them.

7 The range(s) of x for which $x^2 - 3x - 4 > 0$ is (are)

  A $x > 4$ only. B $x < -1$ and $x > 4$. C $-1 < x < 4$.
  D $x < -1$ only. E none of them.

8 In figure 11(a), the shaded area, including the x- and y-axes but not the lines $x = 3$ and $y = 2$, is given by

  A $x \geqslant 0$, $y \geqslant 0$, $y < 2$, $x < 3$.
  B $x \geqslant 0$, $y \leqslant 0$, $y > 2$, $x < 3$.
  C $x \leqslant 0$, $y \geqslant 0$, $y > 2$, $x < 3$.
  D $x \geqslant 0$, $y \geqslant 0$, $y < 2$, $x > 3$.
  E $x \leqslant 0$, $y \geqslant 0$, $y > 2$, $x > 3$.

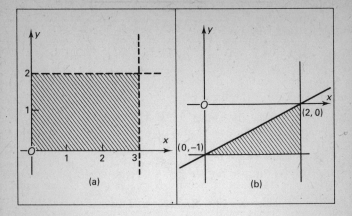

Figure 11

9 In figure 11(b), the shaded area, including all the boundary lines, is given by the inequalities

A $x \leqslant 2$, $y \geqslant -1$, $x-2y \geqslant 2$.
B $x \leqslant 2$, $y \leqslant -1$, $x-2y \leqslant 2$.
C $x \geqslant 2$, $y \leqslant -1$, $x-2y \geqslant 2$.
D $x \leqslant 2$, $y \geqslant -1$, $x-2y \leqslant 2$.
E $x \geqslant 2$, $y \geqslant -1$, $x-2y \geqslant 2$.

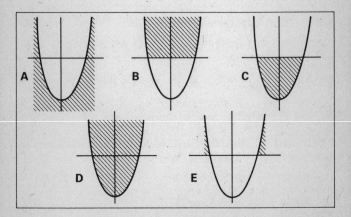

Figure 12

10 Each of the graphs in figure 12 is $y = x^2 - 2$. Which one of the shaded areas represents the inequality $y > x^2 - 2$?

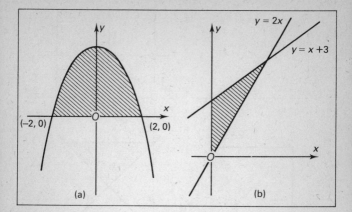

(a)

(b)

Figure 13

11 In figure 13(a), the curve is the graph of $y = 4 - x^2$. The shaded area including the boundaries is the set of points

A $\{(x,y) : y \geqslant 0, \ y \geqslant 4 - x^2\}$.
B $\{(x,y) : y \leqslant 0, \ y \leqslant 4 - x^2\}$.
C $\{(x,y) : x \geqslant 0, \ y \geqslant 4 - x^2\}$.
D $\{(x,y) : x \leqslant 0, \ y \leqslant 4 - x^2\}$.
E $\{(x,y) : y \geqslant 0, \ y \leqslant 4 - x^2\}$.

12 In figure 13(b), $\varepsilon = \{(x,y) : x \text{ real}, \ y \text{ real}\}$,
$P = \{(x,y) : x > 0\}$, $Q = \{(x,y) : y > x + 3\}$,
$R = \{(x,y) : y > 2x\}$.

The shaded area excluding the boundary lines is given by the set

A $P \cap Q' \cap R$ B $P \cap Q \cap R'$ C $P \cap Q' \cap R'$
D $(P \cup Q \cup R)'$ E none of them.

13 In figure 14(a), the area represented by the inequalities
$x > 0 : y < 0$ and $x + y < 7$ is

A j B k C l D m E n

35

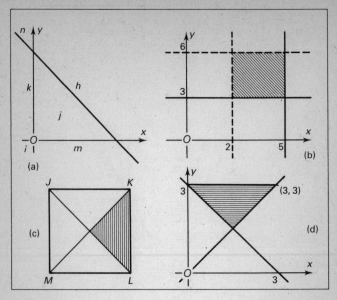

Figure 14

14 In figure 14(b), if $P = \{(x,y) : 2 < x \leqslant 5\}$ and
$Q = \{(x,y) : 3 \leqslant y < 6\}$, the shaded area is represented by the set

A $P \cup Q$ B $P' \cap Q$ C $P \cap Q'$ D $P \cap Q$
E none of them.

15 In figure 14(c), *JKLM* is a square. If $\varepsilon = \{P : P \text{ inside } JKLM\}$,
$X = \{P : PK > PM\}$ and $Y = \{P : PL < PJ\}$, then the shaded
area is

A $X \cap Y$ B $X' \cap Y$ C $(X \cup Y)'$ D $X \cup Y$ E $(X \cap Y)'$

16 In figure 14(d), which pair(s) of inequalities satisfy the shaded
area (excluding the boundary lines)?

1 $y < 3$; $x + y < 3$
2 $y < x$; $x + y > 3$
3 $y < 3$; $y > x$

A 1 only B 2 only C 3 only D 1,2 E 2,3

Chapter 7
Matrices

1 Which of the following matrix products exist?

$$1 \begin{pmatrix} 1 \\ 4 \end{pmatrix}\begin{pmatrix} 3 & 2 \\ 0 & 1 \end{pmatrix} \quad 2 \begin{pmatrix} 4 & 7 \\ 3 & 1 \end{pmatrix}\begin{pmatrix} 1 \\ 5 \end{pmatrix} \quad 3 (5\ 6)\begin{pmatrix} 4 & 8 \\ 3 & 1 \end{pmatrix}$$

A 2 only   B 1,2   C 1,2,3   D 2,3   E 1 only

2 If $P = \begin{pmatrix} 4 & -3 \\ 2 & 1 \end{pmatrix}$ and $Q = \begin{pmatrix} 3 & -1 \\ 0 & 2 \end{pmatrix}$ then PQ is

A $\begin{pmatrix} 12 & 3 \\ 0 & 2 \end{pmatrix}$ B $\begin{pmatrix} 12 & -10 \\ 6 & 0 \end{pmatrix}$ C $\begin{pmatrix} 9 & -10 \\ 7 & 0 \end{pmatrix}$ D $\begin{pmatrix} 12 & -2 \\ 6 & 4 \end{pmatrix}$

E $\begin{pmatrix} 9 & -2 \\ 7 & 4 \end{pmatrix}$

3 If $B = \begin{pmatrix} 3 & 4 \\ -1 & 2 \end{pmatrix}$ then B^2 is

A $\begin{pmatrix} 9 & 16 \\ 1 & 4 \end{pmatrix}$ P $\begin{pmatrix} 5 & 20 \\ -1 & 8 \end{pmatrix}$ C $\begin{pmatrix} 13 & 4 \\ -5 & 8 \end{pmatrix}$ D $\begin{pmatrix} 13 & 4 \\ -5 & 0 \end{pmatrix}$

E $\begin{pmatrix} 5 & 20 \\ -5 & 0 \end{pmatrix}$

4 If $M\begin{pmatrix} 1 \\ 0 \end{pmatrix} = \begin{pmatrix} 5 \\ 2 \end{pmatrix}$ and $M\begin{pmatrix} 1 \\ 2 \end{pmatrix} = \begin{pmatrix} 3 \\ 8 \end{pmatrix}$ and M is a 2×2 matrix, then M is

A $\begin{pmatrix} 5 & -1 \\ 2 & 3 \end{pmatrix}$ B $\begin{pmatrix} 2 & 3 \\ 5 & -1 \end{pmatrix}$ C $\begin{pmatrix} 5 & 2 \\ -1 & 3 \end{pmatrix}$ D $\begin{pmatrix} 5 & -4 \\ 2 & 4 \end{pmatrix}$

E $\begin{pmatrix} 2 & -2 \\ 5 & 4 \end{pmatrix}$

5 If $\begin{pmatrix} 3 & -2 \\ -1 & 4 \end{pmatrix}\begin{pmatrix} 7 & a \\ 2 & -4 \end{pmatrix} = \begin{pmatrix} 17 & 2 \\ 1 & -14 \end{pmatrix}$ then a is

A -1 B -2 C 0   D 2   E 1

6 If $P = \begin{pmatrix} 4 & 2 \\ 3 & -1 \end{pmatrix}$ and $Q = \begin{pmatrix} 1 & -1 \\ 0 & 1 \end{pmatrix}$ then $P - 3Q$ is

A $\begin{pmatrix} 1 & -5 \\ 3 & 4 \end{pmatrix}$ B $\begin{pmatrix} 7 & -5 \\ 3 & -4 \end{pmatrix}$ C $\begin{pmatrix} 1 & -1 \\ 3 & -2 \end{pmatrix}$ D $\begin{pmatrix} 1 & 5 \\ 3 & -4 \end{pmatrix}$

E $\begin{pmatrix} 7 & 3 \\ 3 & -2 \end{pmatrix}$

7 If $P = \begin{pmatrix} 1 & -1 \\ 1 & 0 \end{pmatrix}$ and $Q = \begin{pmatrix} -1 & 1 \\ 0 & 1 \end{pmatrix}$ then $PQ - QP$ is

A $\begin{pmatrix} 0 & 0 \\ 0 & 0 \end{pmatrix}$ B $\begin{pmatrix} -1 & 1 \\ 0 & 1 \end{pmatrix}$ C $\begin{pmatrix} -1 & -1 \\ -2 & 1 \end{pmatrix}$ D $\begin{pmatrix} 1 & 1 \\ 0 & 1 \end{pmatrix}$

E $\begin{pmatrix} 1 & 1 \\ -2 & 1 \end{pmatrix}$

8 A 2×2 matrix is formed by using the digits 2,3,4,5 to form its four elements. The smallest possible value of the determinant of this matrix is

A -7 B -2 C -16 D -14 E 0

9 Which of these matrices is (are) singular?

1 $\begin{pmatrix} 2 & 2 \\ -2 & 2 \end{pmatrix}$ 2 $\begin{pmatrix} 2 & 2 \\ -2 & -2 \end{pmatrix}$ 3 $\begin{pmatrix} -2 & 8 \\ -1 & 4 \end{pmatrix}$

A 1,2   B 2,3   C 1,3   D 1,2,3   E none of them.

10 The inverse of the matrix $\begin{pmatrix} 7 & 5 \\ 3 & 2 \end{pmatrix}$

A $\begin{pmatrix} 2 & -5 \\ -3 & 7 \end{pmatrix}$ B $\begin{pmatrix} -7 & 3 \\ 5 & -2 \end{pmatrix}$ C $\begin{pmatrix} -2 & 5 \\ 3 & -7 \end{pmatrix}$

D $\begin{pmatrix} 7 & -3 \\ -5 & 2 \end{pmatrix}$ E $\begin{pmatrix} -7 & -5 \\ -3 & -2 \end{pmatrix}$

11 If $P = \begin{pmatrix} 8 & 5 \\ 4 & 3 \end{pmatrix}$ then P^{-1} is

A $\begin{pmatrix} 3 & -5 \\ -4 & 8 \end{pmatrix}$ B $\begin{pmatrix} -8 & 4 \\ 5 & -3 \end{pmatrix}$ C $\frac{1}{4}\begin{pmatrix} -8 & 4 \\ 5 & -3 \end{pmatrix}$

D $\frac{1}{4}\begin{pmatrix} 3 & -5 \\ -4 & 8 \end{pmatrix}$ E $\frac{1}{4}\begin{pmatrix} 3 & -4 \\ -5 & 8 \end{pmatrix}$

12 Given that $(3 \ a)\begin{pmatrix} -2 \\ 4 \end{pmatrix} = (14)$ the value of a is

  A 4 B 1 C 5 D 2 E -2

13 If $\begin{pmatrix} x \\ y \end{pmatrix} + \begin{pmatrix} y \\ 2 \end{pmatrix} = \begin{pmatrix} -5 \\ -3 \end{pmatrix}$ then $\begin{pmatrix} x \\ y \end{pmatrix}$ is

  A $\begin{pmatrix} -10 \\ -5 \end{pmatrix}$ B $\begin{pmatrix} -6 \\ -1 \end{pmatrix}$ C $\begin{pmatrix} 0 \\ 5 \end{pmatrix}$ D $\begin{pmatrix} -8 \\ 2 \end{pmatrix}$

  E $\begin{pmatrix} 0 \\ -5 \end{pmatrix}$

14 How many values of the matrix $\begin{pmatrix} x \\ y \end{pmatrix}$ satisfy this equation?

$$\begin{pmatrix} 3 & 4 \\ -1 & 0 \end{pmatrix}\begin{pmatrix} x \\ y \end{pmatrix} = \begin{pmatrix} -3 \\ -1 \end{pmatrix}$$

  A none B one C two D three E an infinite number

15 If $\begin{pmatrix} 3 & 2 \\ 5 & 2 \end{pmatrix}\begin{pmatrix} x \\ y \end{pmatrix} = \begin{pmatrix} 5 \\ 11 \end{pmatrix}$ then $\begin{pmatrix} x \\ y \end{pmatrix}$ is

  A $\begin{pmatrix} 2 \\ -\frac{1}{2} \end{pmatrix}$ B $\begin{pmatrix} -2 \\ 5\frac{1}{2} \end{pmatrix}$ C $\begin{pmatrix} -3 \\ 7 \end{pmatrix}$ D $\begin{pmatrix} 3 \\ -2 \end{pmatrix}$

  E $\begin{pmatrix} -8 \\ -4\frac{1}{2} \end{pmatrix}$

16 If $\begin{pmatrix} a & 5 \\ -2 & b \end{pmatrix}\begin{pmatrix} 2 \\ -1 \end{pmatrix} = \begin{pmatrix} 9 \\ 0 \end{pmatrix}$ then $a-b$ is

  A 3 B 11 C -3 D -11 E none of them.

17 If $\begin{pmatrix} 5 & 2h \\ h & 3 \end{pmatrix}\begin{pmatrix} 3 \\ 1 \end{pmatrix} = \begin{pmatrix} k \\ 9 \end{pmatrix}$ the value of k is

  A 7 B 11 C 19 D 21 E 9

18 P is a matrix such that $\begin{pmatrix} 8 & 6 \\ 4 & 3 \end{pmatrix} - 2P = \begin{pmatrix} 2 & 2 \\ -4 & -7 \end{pmatrix}$. P is

  A $\begin{pmatrix} 3 & 2 \\ 4 & 5 \end{pmatrix}$ B $\begin{pmatrix} 5 & 2 \\ 4 & 5 \end{pmatrix}$ C $\begin{pmatrix} 3 & 4 \\ 4 & 5 \end{pmatrix}$ D $\begin{pmatrix} 5 & 2 \\ 0 & -2 \end{pmatrix}$

  E $\begin{pmatrix} 5 & 4 \\ 4 & -2 \end{pmatrix}$

19 The matrix $P = \begin{pmatrix} 5 & 0 \\ 0 & 2 \end{pmatrix}$ and Q is of the form $\begin{pmatrix} a & b \\ 0 & c \end{pmatrix}$ and $PQ = P + Q$, Q is the matrix

A $\begin{pmatrix} \frac{5}{6} & 5 \\ 0 & 2 \end{pmatrix}$ B $\begin{pmatrix} \frac{1}{4} & 0 \\ 0 & 1 \end{pmatrix}$ C $\begin{pmatrix} \frac{5}{6} & 5 \\ 0 & 1 \end{pmatrix}$ D $\begin{pmatrix} \frac{5}{4} & 0 \\ 0 & 2 \end{pmatrix}$

E $\begin{pmatrix} \frac{1}{2} & 4 \\ 0 & 3 \end{pmatrix}$

20 If $P = \begin{pmatrix} 1 & -2 \\ 0 & 3 \end{pmatrix}$ and $Q = \begin{pmatrix} 4 & -2 \\ 0 & 6 \end{pmatrix}$ are matrices such that $P^2 + Q = kP$, then the value of k is

A 1   B 2   C 3   D 4   E 5

Chapter 8
Elementary Geometry

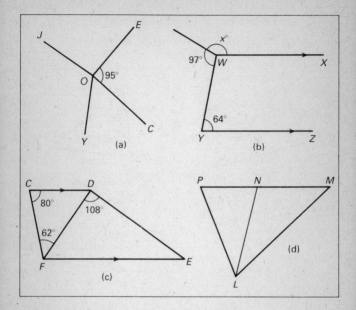

Figure 15

1 In figure 15(a), angle JOY = angle JOE = 2 × angle YOC. Angle YOC is

  A 53°   B 45°   C 16°   D 72°   E 24°

2 In figure 15(b), WX and YZ are parallel. The value of $x°$ is

  A 141   B 147   C 131   D 137   E 151

3 In figure 15(c), CD is parallel to EF. Angle DEF is

  A 80°   B 44°   C 34°   D 38°   E 62°

4 In figure 15(d), LM = MP and LN = LP. If angle LMP = 50°, the value of angle MLN is

  A 25°   B 50°   C 15°   D 35°   E 40°

5 The sizes of the angles in a triangle are in the ratio $3 : 4 : 5$. The largest angle is

    A 15°   B 45°   C 90°   D 75°   E 105°

6 A triangle is right angled. The three angles are $a°$, $(a+30)°$ and $(b-20)°$. Which one of the following must be false?

    A $a=40$ B $a=60$ C $a=90$ D $b=40$ E $b=110$

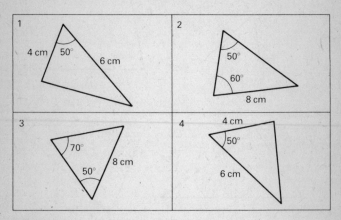

Figure 16

7 In figure 16, which of the following pairs of triangles must be congruent?

    A 1,4 only   B 1,2 and 3,4   C 2,4 only   D 2,3 only
    E 1,4 and 2,3

8 In figure 17(a), angle YXZ = angle PQR and $XY = PQ$. The triangles must be congruent if, further,

    A angle XZY = angle QPR.   B $XZ = PR$.   C $YZ = QR$.
    D $XZ = QR$.   E none of them.

9 In figure 17(b), if angle PRQ = angle PST and angle PQR = angle PTS, then

    A $\dfrac{PQ}{PS} = \dfrac{PR}{PT}$ B $\dfrac{PQ}{PT} = \dfrac{PR}{PS}$ C $\dfrac{PQ}{QS} = \dfrac{PR}{RT}$ D $\dfrac{PQ}{PS} = \dfrac{PR}{ST}$
    E none of them.

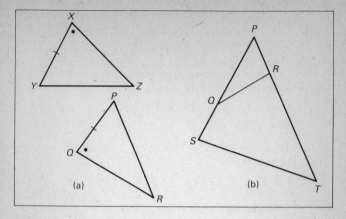

Figure 17

10 In figure 18(a), the triangles *JON* and *LIZ* have sides with lengths as shown. Angles *OJN* and *LZI* are equal. Angles *LIZ* and *ONJ* are equal. The length of *IZ*, in cm, is

A 10 B 6·6 C 4·8 D 9·6 E 16

11 In figure 18(b) *DE* is parallel to *AC*. *AD* = 3 cm, *DB* = 2 cm, *DE* = 4 cm. *AC* is

A 6 cm B 10 cm C 1·6 cm D 9 cm E 12 cm

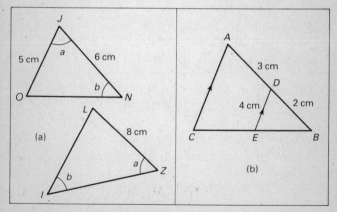

Figure 18

12 Triangles ABC and XYZ are similar. AB corresponds to XY and $AB = 4$ cm, $XY = 7$ cm. Which of the following statements must be true?

1 Angle $XZY =$ angle ACB.
2 The areas of the triangles are in the ratio 16 : 49.
3 Angle $BAC =$ angle YXZ.

A 1,2   B 2,3   C 1,3   D 1,2,3
E none of them.

13 The ratio of the heights of two similar cones is 1 : 2. The ratio of their volumes is

A 1 : 2   B 1 : 4   C 1 : 6   D 1 : 8   E 3 : 8

14 A and B are similar solids. Their surface areas are 36 cm^2 and 100 cm^2 respectively. The height of B is 15 cm. The height of A, in cm, is

A 1·8   B 5·4   C 9   D 12   E 25

15 The lengths of the sides of three triangles are

1 8 cm, 15 cm, 17 cm
2 11 cm, 14 cm, 20 cm
3 15 cm, 20 cm, 25 cm

Which of the triangles is (are) right angled?

A 1 only   B 2 only   C 3 only   D 1,3   E 2,3

16 A rectangular box has length 4 cm, breadth 3 cm and height 5 cm. The length of the diagonal of the box, in cm, is

A $\sqrt{50}$ B 10   C $\sqrt{34}$ D $\sqrt{41}$ E 5

17 In figure 19(a), the value of x, in cm, is

A $\sqrt{41}$ B 4   C $\sqrt{146}$ D 5   E $\sqrt{23}$

18 In figure 19(b), the value of $x + y + z$ is

A 180°   B 150°   C 330°   D 360°   E 400°

19 The sum of the interior angles of an 11 sided polygon is

A 11 right angles.   B 1620°.   C 1500°.   D 1100°.
E none of them.

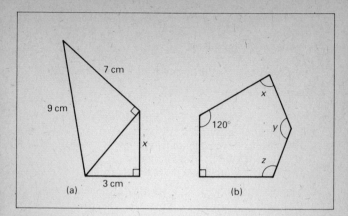

Figure 19

20 Each interior angle of a regular 9 sided figure is

    A 160° B 144° C 150° D 130° E 140°

21 The number of axes of symmetry of a regular octagon is

    A 16 B 8 C 4 D 2 E 0

22 *PQRSTU* is a regular hexagon. The length of *PS* is 8 cm. The length of *QR*, in cm, is

    A 4 B 8 C $\sqrt{8}$ D 6 E $2\sqrt{8}$

23 *PQRS* is a parallelogram. *P* is the point (1,1), *Q* is (3,4) and *S* is (6,2).

    The co-ordinates of *R* are

    A $(-2,3)$ B $(4,-1)$ C $(8,5)$ D $(3\frac{1}{2},1\frac{1}{2})$
    E $(2,2\frac{1}{2})$

24 The diagonals of a quadrilateral are at right angles. The quadrilateral could be a

    A rectangle. B trapezium. C parallelogram.
    D rhombus. E pentagon.

25 Which of the following properties is (are) true of any parallelogram?
    1 It has rotational symmetry.
    2 It has one axis of symmetry.
    3 The diagonals bisect the angles of the parallelogram.

    A 1 only B 2 only C 3 only D 2,3 E 1,2

26 Which of the following properties is (are) common to a rhombus and a rectangle?
    1 The diagonals are equal in length.
    2 The diagonals bisect at right angles.
    3 The four sides are equal in length.

    A 1 only B 2 only C 2,3 D 1,2 E none of them.

27 The side of a rhombus is 13 cm in length. The longer diagonal is 24 cm. The length of the shorter diagonal, in cm, is

    A 8 B 10 C 12 D 14 E 16

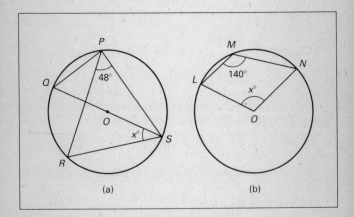

(a)   (b)

Figure 20

28 In figure 20(a), QS is a diameter of the circle centre O. $x° =$

    A 42 B 48 C 52 D 58 E 32

29 In figure 20(b), O is the centre of the circle. $x°$ is

    A 140 B 70 C 280 D 80 E 40

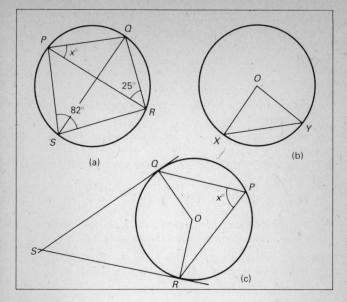

Figure 21

30 In figure 21(a), *PQRS* is a cyclic quadrilateral. $x°$ is

A 73   B 57   C 37   D 83   E 47

31 In figure 21(b), the chord *XY* is equal in length to the radius.
The ratio of the length of the minor arc *XY* to the length of
the major arc *XY* is

A 1 : 6   B 1 : 5   C 2 : 5   D 1 : 3   E 1 : 8

32 In figure 21(c), *QS* and *RS* are the tangents to the circle at *Q*
and *R*. Angle *QSR*, in degrees, is

A $2x$ B x C $180 - x$ D $180 - 2x$ E $180 - \dfrac{x}{2}$

33 The length of the tangent from a point *T* to a given circle is
8 cm. *T* is 10 cm from the centre. The diameter, in cm, is

A 6   B 8   C 10   D 12   E 14

Chapter 9
Transformation Geometry

1 The point $P(4,2)$ is reflected in the line $x=0$. The co-ordinates of the image P' are

  A $(-4,2)$ B $(4,-2)$ C $(-4,-2)$ D $(4,0)$ E $(0,4)$

2 The point $P(-5,2)$ is reflected in the line $y=-x$. The image P' is the point

  A $(5,2)$ B $(5,-2)$ C $(2,5)$ D $(-2,5)$ E $(-5,-2)$

3 A triangle ABC with vertices $(0,0)$, $(4,2)$, $(2,-3)$, is reflected in the x-axis. The co-ordinates of the image $A'B'C'$ are

  A $(0,0)$, $(-4,2)$, $(-2,-3)$. B $(0,0)$, $(4,-2)$, $(2,3)$.
  C $(0,0)$, $(-4,2)$, $(2,3)$. D $(1,1)$, $(4,-2)$, $(2,-3)$.
  E $(0,0)$, $(-4,-2)$, $(-2,-3)$.

4 The straight line $y=5-3x$ is reflected in the line $x=2$. The invariant point is

  A $(2,1)$ B $(2,-1)$ C $(0,0)$ D $(1,2)$ E $(-1,2)$

5 The point $A(-3,4)$ is given a translation 5 units parallel with the $+x$-axis and 2 units parallel with the $-y$-axis. The co-ordinates of the image A' are

  A $(2,2)$ B $(-8,6)$ C $(2,6)$ D $(-8,-2)$
  E $(-2,2)$

6 The points on the straight line $y=x+3$ are given a translation $\begin{pmatrix} -2 \\ 4 \end{pmatrix}$. The images lie on the straight line

  A $y=9-x$ B $y=x+9$ C $y=x+3$ D $y=3-x$
  E $y=-9-x$

7 The point $A(3,-2)$ is rotated about $(0,0)$ through $90°$ anti-clockwise. The image A' is

  A $(-3,-2)$ B $(2,-3)$ C $(-2,-3)$ D $(-3,2)$
  E $(2,3)$

8 The straight line with co-ordinates $P(-4,-2)$, $Q(-2,3)$ is rotated through 90° clockwise about (0,0). The co-ordinates of $P'Q'$ are

A $(-2,4)$, $(2,3)$.   B $(-2,4)$, $(3,-2)$.   C $(-4,2)$, $(3,2)$.
D $(-2,4)$, $(3,2)$.   E $(-4,2)$, $(2,3)$.

9 The point $A(4,6)$ is given an enlargement scale factor $-1\cdot5$ in the origin. The co-ordinates of image A' are

A $(6,9)$ B $(-6,-9)$ C $(-3,-4)$ D $(-2,-3)$ E $(6,-9)$

10 A triangle PQR is enlarged by a scale factor 4 in centre (0,0). The image is triangle $P'Q'R'$. Which of the following statements is (are) true?
1 The perimeter of the triangle $PQR = \frac{1}{4}$ of the perimeter of triangle $P'Q'R'$.
2 The area of triangle $PQR = \frac{1}{4}$ of the area of triangle $P'Q'R'$.
3 Angle PQR $= \frac{1}{4}$ of angle $P'Q'R'$.

A 1 only   B 2 only   C 3 only   D 1,2   E 2,3

11 A parallelogram with an area of 9 cm^2 is given an enlargement of scale factor 4 in (0,0). The area of the image, in cm^2, is

A 36   B 144   C 6   D 72   E none of them.

12. M is a reflection in the line $x=0$; R is a rotation of $+90°$ in (0,0); P is the point (3,2). The point $MR(P)$ is

A $(3,-2)$.   B $(-2,-3)$.   C $(-2,3)$.   D $(2,3)$.
E $(-3,-2)$.

13 M is a reflection in the line $y=x$; H is a half turn about (0,0) and T is a translation $\begin{pmatrix} 2 \\ 1 \end{pmatrix}$.

If Q is the point (4,1), then $THM(Q)$ is the point

A $(0,4)$ B $(-4,-5)$ C $(-2,-6)$ D $(-6,-2)$ E $(1,-3)$

14 If M is a reflection in the x-axis and N is a reflection in the y-axis, then the transformation MN is equivalent to

A a reflection in $y=x$.   B a reflection in $y=-x$.
C an anticlockwise rotation of 90° about (0,0).
D a clockwise rotation of 90° about (0,0).
E a rotation of 180° about (0,0).

15 If M is a reflection in $x=0$ and H is a half turn about the origin, which of the following statements is (are) true?

   1 $H^2 = I$ 2 $M^2 = H^2$ 3 $M^2H = H^2M$

   A 1 only   B 2 only   C 3 only   D 1 and 2   E 1, 2 and 3

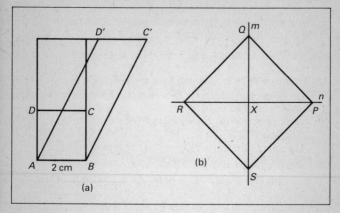

(a)   (b)

Figure 22

16 In figure 22(a), $ABCD$ is a square which is transformed onto $ABC'D'$ by a stretch of 3 units parallel to the $+y$-axis, followed by a shear of 3 units parallel to the $+x$-axis. The area of $ABC'D'$, in square units, is

   A 6   B 8   C 10   D 12   E 14

17 In figure 22(b), M is a reflection in the line m. N is a reflection in the line n. H is a $\frac{1}{2}$ turn about X. Which of the following statements is (are) true?

   1 $MN = H$
   2 $MH = HM$
   3 $MH = N$

   A 1 only   B 2 only   C 2,3   D 1,3   E 1,2,3

Chapter 10
Transformation Matrices

1 The transformation $\begin{pmatrix} 2 & 2\cdot5 \\ 0 & -2 \end{pmatrix}$

transforms the point $(2, -1)$ onto

A $(6\cdot5,0)$ B $(6\cdot5,2)$ C $(1\cdot5,-2)$ D $(1\cdot5,2)$
E $(1\cdot5,0)$

2 The transformation $\begin{pmatrix} 4 & 5 \\ 2 & 3 \end{pmatrix}$ maps the point A onto the point
A'. The transformation which maps A' onto A is

A $\begin{pmatrix} 3 & 2 \\ 5 & 4 \end{pmatrix}$ B $\frac{1}{2}\begin{pmatrix} 3 & -5 \\ -2 & 4 \end{pmatrix}$ C $\begin{pmatrix} 2 & 3 \\ 4 & 5 \end{pmatrix}$ D $\frac{1}{2}\begin{pmatrix} -4 & 2 \\ 5 & -3 \end{pmatrix}$
E none of them.

3 Under a transformation $(1,0) \rightarrow (5,7)$ and $(0,1) \rightarrow (4,3)$. The
transformation matrix is

A $\begin{pmatrix} 5 & 4 \\ 7 & 3 \end{pmatrix}$ B $\begin{pmatrix} 7 & 3 \\ 5 & 4 \end{pmatrix}$ C $\begin{pmatrix} 5 & 7 \\ 4 & 3 \end{pmatrix}$ D $\begin{pmatrix} 4 & 3 \\ 5 & 7 \end{pmatrix}$
E none of them.

4 The transformation $\begin{pmatrix} 4 & -2 \\ 1 & -2 \end{pmatrix}$ maps the point $(4,k)$ onto $(4,l)$.
The value of l is

A -2 B -4 C -6 D -8 E 0

5 The transformation represented by the matrix $\begin{pmatrix} 0 & -1 \\ -1 & 0 \end{pmatrix}$ is

A a rotation of $+90°$ about $(0,0)$.
B a reflection in the line $y=0$.
C a reflection in the line $y=-x$.
D a half turn about $(0,0)$.
E a rotation of $-90°$ about $(0,0)$.

6 Which of the following transformations represents a half turn
in $(0,0)$?

A $\begin{pmatrix} 0 & -1 \\ -1 & 0 \end{pmatrix}$ B $\begin{pmatrix} -1 & 0 \\ 0 & 1 \end{pmatrix}$ C $\begin{pmatrix} 0 & -1 \\ 1 & 0 \end{pmatrix}$

D $\begin{pmatrix} -1 & 0 \\ 0 & -1 \end{pmatrix}$ E $\begin{pmatrix} 1 & 0 \\ 0 & -1 \end{pmatrix}$

7 The transformation $\begin{pmatrix} 2 & 0 \\ 0 & 2 \end{pmatrix}$ maps the triangle ABC onto triangle $A'B'C'$. The area of triangle $A'B'C'$ is

A halved.   B doubled.   C four times larger.
D unchanged.   E three times larger.

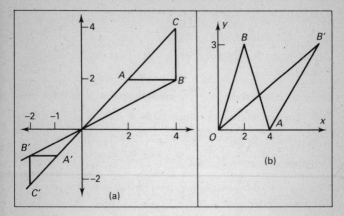

Figure 23

8 In figure 23(a), the matrix which transforms triangle ABC onto triangle $A'B'C'$ is

A $\begin{pmatrix} -2 & 0 \\ 0 & -2 \end{pmatrix}$ B $\begin{pmatrix} \frac{1}{2} & 0 \\ 0 & \frac{1}{2} \end{pmatrix}$ C $\begin{pmatrix} 0 & -2 \\ -2 & 0 \end{pmatrix}$

D $\begin{pmatrix} -\frac{1}{2} & 0 \\ 0 & -\frac{1}{2} \end{pmatrix}$ E none of them.

9 In figure 23(b), triangle OAB with vertices $O(0,0)$, $A(4,0)$, $B(2,3)$, is sheared parallel to the x-axis onto triangle OAB' so that $B \to B'(8,3)$.
The transformation matrix is

A $\begin{pmatrix} 1 & 0 \\ 2 & 1 \end{pmatrix}$ B $\begin{pmatrix} 1 & 1 \\ 0 & 2 \end{pmatrix}$ C $\begin{pmatrix} 2 & 2 \\ 0 & 1 \end{pmatrix}$ D $\begin{pmatrix} 2 & 1 \\ 0 & 1 \end{pmatrix}$

E $\begin{pmatrix} 1 & 2 \\ 0 & 1 \end{pmatrix}$

10 If $P = \begin{pmatrix} 0 & 1 \\ 1 & 0 \end{pmatrix}$ and $Q = \begin{pmatrix} 0 & -1 \\ -1 & 0 \end{pmatrix}$, which of the following statements is (are) true?

1 P is the inverse of Q.
2 Q represents a reflection in the line $x + y = 0$.
3 PQ is equivalent to a rotation of $180°$.

A 1 only   B 2 only   C 2,3   D 1,2,3   E 1,2

11 U and V are the transformations

$$U : \begin{pmatrix} x \\ y \end{pmatrix} \rightarrow \begin{pmatrix} 1 & 2 \\ 0 & 1 \end{pmatrix} \begin{pmatrix} x \\ y \end{pmatrix} \quad V : \begin{pmatrix} x \\ y \end{pmatrix} \rightarrow \begin{pmatrix} x \\ y \end{pmatrix} + \begin{pmatrix} 1 \\ 2 \end{pmatrix}$$

The image of $(2, -1)$ under the transformation UV is

A $(-1, 2)$ B $(5, 1)$ C $(8, 4)$ D $(1, 1)$ E $(1, 3)$

12 $M = \begin{pmatrix} 1 & 0 \\ 0 & -1 \end{pmatrix}$ $T = \begin{pmatrix} 0 & 1 \\ -1 & 0 \end{pmatrix}$ $P = \begin{pmatrix} 0 & 1 \\ 1 & 0 \end{pmatrix}$ $I = \begin{pmatrix} 1 & 0 \\ 0 & 1 \end{pmatrix}$

Which of the following statements is (are) true?

1 M is a reflection in the x-axis.
2 $MP = PM$.
3 $(MT)P = M(TP) = I$.

A 1 only   B 1 and 2   C 2 and 3   D 1 and 3
E 1,2 and 3

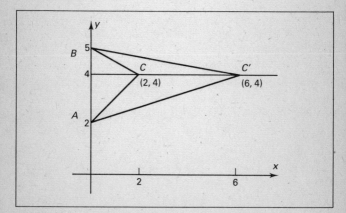

Figure 24

13 In figure 24, the triangle ABC is stretched parallel to the x-axis so that $C(2,4) \rightarrow C'(6,4)$. The transformation matrix is

A $\begin{pmatrix} 3 & 0 \\ 0 & 1 \end{pmatrix}$ B $\begin{pmatrix} 1 & 0 \\ 0 & 3 \end{pmatrix}$ C $\begin{pmatrix} 1 & 3 \\ 0 & 1 \end{pmatrix}$ D $\begin{pmatrix} 1 & 1 \\ 3 & 0 \end{pmatrix}$

E $\begin{pmatrix} 0 & 3 \\ 1 & 0 \end{pmatrix}$

Chapter 11
Vectors

1 **a** and **b** are two vectors. Which of the diagrams in figure 25(a) represent the condition that **a** = **b**?

  A 1 only B 2 only C 1 and 2 D 3 and 4 E 4 only

2 If $\overrightarrow{OA} = \begin{pmatrix} -5 \\ 12 \end{pmatrix}$ then the length of $\overrightarrow{OA}$ is

  A 7 units. B −7 units. C −13 units. D 13 units.
  E 17 units.

3 If $\mathbf{a} = \begin{pmatrix} 5 \\ 3 \end{pmatrix}$ then | **a** | is

  A 4 units. B 2 units. C $\sqrt{34}$ units. D 8 units.
  E $\sqrt{8}$ units.

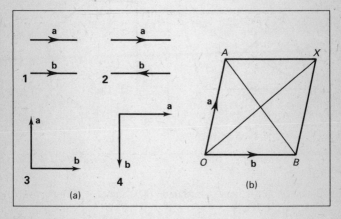

(a)

Figure 25

4 In figure 25(b), $OAXB$ is a parallelogram. $\overrightarrow{OA} = \mathbf{a}$ and $\overrightarrow{OB} = \mathbf{b}$. Which of the following statements is (are) true?

  1 $\overrightarrow{OX} = \mathbf{a} + \mathbf{b}$ 2 $\overrightarrow{AB} = \mathbf{a} - \mathbf{b}$
  3 if | **a** | = | **b** |, then $\overrightarrow{AB}$ and $\overrightarrow{OX}$ are perpendicular vectors.

  A 1 only B 1,2 C 2,3 D 1,3 E none of them.

5 *OABC* is a parallelogram. Vectors **p** and **q** represent two of its adjacent sides. The vector for one of the diagonals of the parallelogram is

A 2**p** B 2**q** C $\frac{1}{2}$**p** D $\frac{1}{2}$**q** E **p**+**q**

6 In figure 26(a), *OACB* is a parallelogram. $\overrightarrow{XC}$ is

A $\frac{1}{2}$**b**+**a** B $\frac{1}{2}$(**b**−**a**) C $\frac{1}{2}$(**a**+**b**) D $\frac{1}{2}$(**a**−**b**) E $\frac{1}{2}$**a**+**b**

7 In the same figure, 26(a), $\overrightarrow{AX}$ is

A $\frac{1}{2}$**b**−**a** B $\frac{1}{2}$**a**−**b** C $\frac{1}{2}$(**a**+**b**) D $\frac{1}{2}$(**a**−**b**) E $\frac{1}{2}$(**b**−**a**)

8 In figure 26(b), the vector $\overrightarrow{PQ}$ is

A **a**+**b**+**c**+**d** B **a**−**b**−**c**−**d** C **a**+**b**−**c**−**d**
D **b**+**d**−**c**−**a** E **b**+**d**+**c**−**a**

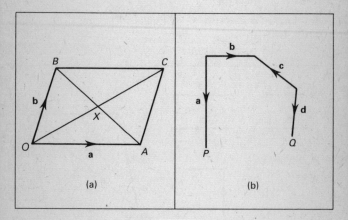

(a)

(b)

Figure 26

9 Points *X* and *Y* have co-ordinates (4,3) and (2, −1) respectively. The vector $\overrightarrow{YX}$ is

A $\begin{pmatrix} -6 \\ 2 \end{pmatrix}$ B $\begin{pmatrix} 6 \\ 2 \end{pmatrix}$ C $\begin{pmatrix} 2 \\ 6 \end{pmatrix}$ D $\begin{pmatrix} 2 \\ 4 \end{pmatrix}$ E $\begin{pmatrix} -2 \\ 4 \end{pmatrix}$

10 $\overrightarrow{OQ}$ is $\begin{pmatrix} 1 \\ 2 \end{pmatrix}$ and $\overrightarrow{OP}$ is $\begin{pmatrix} 5 \\ 8 \end{pmatrix}$. The vector $\overrightarrow{PQ}$ is

A $\begin{pmatrix} -4 \\ 6 \end{pmatrix}$ B $\begin{pmatrix} 4 \\ 6 \end{pmatrix}$ C $\begin{pmatrix} -4 \\ -6 \end{pmatrix}$ D $\begin{pmatrix} 6 \\ 10 \end{pmatrix}$ E $\begin{pmatrix} -6 \\ 10 \end{pmatrix}$

11 If $\mathbf{a} = \begin{pmatrix} 3 \\ -2 \end{pmatrix}$ and $\mathbf{b} = \begin{pmatrix} 4 \\ -3 \end{pmatrix}$ which of the following is (are) true?

1 $4\mathbf{a} = \begin{pmatrix} 12 \\ -8 \end{pmatrix}$ 2 $|\mathbf{b}| = 5$ 3 $2\mathbf{a} + 3\mathbf{b} = \begin{pmatrix} 18 \\ -13 \end{pmatrix}$

A 1 only   B 2 only   C 3 only   D 2,3
E 1,2,3

12 Given that $\mathbf{a} = \begin{pmatrix} -5 \\ 5 \end{pmatrix}$ and $\mathbf{b} = \begin{pmatrix} 3 \\ -2 \end{pmatrix}$ and $\mathbf{c}$ is a vector such that $\mathbf{a} + \mathbf{b} + \mathbf{c} = 0$, then $\mathbf{c}$ is

A $\begin{pmatrix} 2 \\ 3 \end{pmatrix}$ B $\begin{pmatrix} -2 \\ -3 \end{pmatrix}$ C $\begin{pmatrix} 8 \\ 7 \end{pmatrix}$ D $\begin{pmatrix} 2 \\ -3 \end{pmatrix}$ E $\begin{pmatrix} -8 \\ -7 \end{pmatrix}$

13 If $\overrightarrow{AO} = \mathbf{a}$ and $\overrightarrow{OB} = \mathbf{b}$ and $\overrightarrow{AB} = \mathbf{x}$ then $\mathbf{x}$ is

A $\mathbf{a} + \mathbf{b}$ B $\mathbf{a} - \mathbf{b}$ C $\mathbf{b} - \mathbf{a}$ D $-\mathbf{a} - \mathbf{b}$
E none of them.

14 In figure 27(a), the position vectors of A and B are $\mathbf{a}$ and $\mathbf{b}$ respectively. M and N are the midpoints of OA and OB respectively. The vector MN is

A $\frac{1}{2}(\mathbf{a} - \mathbf{b})$ B $\frac{1}{2}(\mathbf{b} - \mathbf{a})$ C $\frac{1}{2}(\mathbf{a} + \mathbf{b})$ D $-\frac{1}{2}\mathbf{a} - \frac{1}{2}\mathbf{b}$
E none of them.

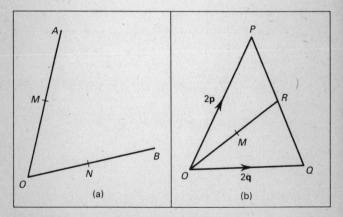

Figure 27

15 The position vectors of the points A and B are
$\overrightarrow{OA} = \begin{pmatrix} -2 \\ 4 \end{pmatrix}$ and $\overrightarrow{OB} = \begin{pmatrix} 5 \\ 3 \end{pmatrix}$. The vector $\overrightarrow{AB}$ is

A $\begin{pmatrix} -7 \\ -1 \end{pmatrix}$ B $\begin{pmatrix} -7 \\ 1 \end{pmatrix}$ C $\begin{pmatrix} 3 \\ 7 \end{pmatrix}$ D $\begin{pmatrix} 7 \\ -1 \end{pmatrix}$
E $\begin{pmatrix} -3 \\ -7 \end{pmatrix}$

16 In figure 27(b), $\overrightarrow{OP} = 2\mathbf{p}$ and $\overrightarrow{OQ} = 2\mathbf{q}$; R is the midpoint of PQ; M is the midpoint of OR. The vector $\overrightarrow{MR}$ is

A $\frac{1}{2}(\mathbf{p} - \mathbf{q})$ B $\frac{1}{2}(\mathbf{p} + \mathbf{q})$ C $\mathbf{p} + \mathbf{q}$ D $\mathbf{p} - \mathbf{q}$ E $2(\mathbf{p} - \mathbf{q})$

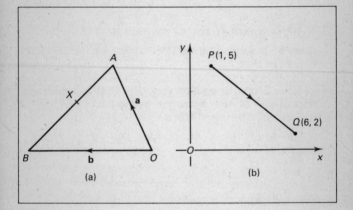

Figure 28

17 In figure 28(a), $\overrightarrow{OA} = \mathbf{a}$ and $\overrightarrow{OB} = \mathbf{b}$. $\overrightarrow{AX} : \overrightarrow{XB} = 3 : 4$.
$\overrightarrow{AX}$ is

A $\frac{3}{4}(\mathbf{b} - \mathbf{a})$ B $\frac{3}{4}(\mathbf{a} - \mathbf{b})$ C $\frac{3}{7}(\mathbf{a} - \mathbf{b})$ D $\frac{3}{7}(\mathbf{b} - \mathbf{a})$
E $\frac{3}{7}(\mathbf{a} + \mathbf{b})$

18 In figure 28(b), $\overrightarrow{PQ}$ is

A $\begin{pmatrix} 5 \\ 3 \end{pmatrix}$ B $\begin{pmatrix} -3 \\ -5 \end{pmatrix}$ C $\begin{pmatrix} 3 \\ -5 \end{pmatrix}$ D $\begin{pmatrix} -5 \\ 3 \end{pmatrix}$
E $\begin{pmatrix} 5 \\ -3 \end{pmatrix}$

19 If $\mathbf{a} = \begin{pmatrix} 5 \\ 6 \end{pmatrix}$ and $\mathbf{b} = \begin{pmatrix} 4 \\ 2 \end{pmatrix}$ then the vector $\begin{pmatrix} -2 \\ 6 \end{pmatrix}$ is

A $2\mathbf{a} - 3\mathbf{b}$ B $3\mathbf{a} - 2\mathbf{b}$ C $2\mathbf{a} + \mathbf{b}$ D $\mathbf{a} + 2\mathbf{b}$
E $2\mathbf{b} - 3\mathbf{a}$

20 The angle between two vectors $\mathbf{a}$ and $\mathbf{b}$ is 60°. If $|\mathbf{a}| = |\mathbf{b}| = 1$, then $|\mathbf{a} - \mathbf{b}|$ is

A $\sqrt{2}$ B $\sqrt{3}$ C 1 D 2 E none of them.

21 In a triangle PQR, $\overrightarrow{PQ} = 3\mathbf{i} + 2\mathbf{j}$ and $\overrightarrow{QR} = 4\mathbf{i} - 5\mathbf{j}$.
$\overrightarrow{PR}$ is

A $\mathbf{i} + 3\mathbf{j}$ B $-\mathbf{i} - 2\mathbf{j}$ C $7\mathbf{i} + 3\mathbf{j}$ D $-7\mathbf{i} - 3\mathbf{j}$
E $7\mathbf{i} - 3\mathbf{j}$

22 In a triangle ABC, $\overrightarrow{AB} = 4\mathbf{i} - 3\mathbf{j}$; $\overrightarrow{AC} = 2\mathbf{i} + 3\mathbf{j}$. $\overrightarrow{BC}$ is

A $-2\mathbf{i} - 6\mathbf{j}$ B $-2\mathbf{i} + 6\mathbf{j}$ C $2\mathbf{i}$ D $-2\mathbf{i}$ E $6\mathbf{i}$

23 In figure 29, an aircraft flies on a course due east represented by the vector $\mathbf{c}$. The wind is blowing from the northwest, represented by the vector $\mathbf{w}$. The velocity triangle showing the resultant speed and direction, vector $\mathbf{t}$, is given by

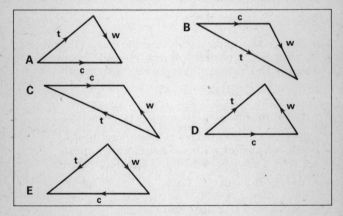

Figure 29

Chapter 12
Basic Arithmetic

1 If $p = \frac{5}{7}$, $q = \frac{4}{5}$, $r = \frac{7}{10}$, then

  A $p > q > r$ B $q > p > r$ C $p < q < r$ D $q < p < r$
  E $q > r > p$

2 The value of $\frac{5}{6} + \frac{4}{9}$ is

  A $\frac{9}{15}$ B $\frac{23}{18}$ C $\frac{9}{54}$ D $\frac{20}{54}$ E $\frac{8}{27}$

3 The value of $(\frac{1}{6} + \frac{4}{5}) + \frac{3}{10}$ is

  A $\frac{14}{30}$ B $\frac{7}{40}$ C $\frac{7}{30}$ D $\frac{7}{300}$ E $\frac{13}{30}$

4 $\frac{9}{20}$ is the result of evaluating

  A $\frac{5}{11} + \frac{4}{9}$ B $\frac{3}{5} \div 1\frac{1}{3}$ C $\frac{3}{5}$ of $1\frac{1}{4}$ D $\frac{2}{11} \times \frac{7}{9}$
  E $\frac{10}{23} - \frac{1}{3}$

5 The average of $\frac{1}{8}$ and $\frac{1}{16}$ is

  A $\frac{3}{16}$ B $\frac{1}{12}$ C $\frac{1}{24}$ D $\frac{3}{32}$ E $\frac{3}{8}$

6 In an election, $\frac{3}{8}$ of the eligible voters do not cast their vote. Of
  the votes cast, A receives $\frac{2}{5}$, B receives $\frac{1}{3}$ and C receives the remainder. The fraction of the total eligible vote received by C is

  A $\frac{11}{15}$ B $\frac{1}{6}$ C $\frac{31}{40}$ D $\frac{1}{4}$ E $\frac{5}{24}$

7 Which of the following statements is (are) true?

  1 0·036 is a number written to 2 significant figures.
  2 6700 is a number written to 2 significant figures.
  3 1·0063 is a number written to 3 significant figures.

  A 1 only B 2 only C 3 only D 1,2 E 2,3

8 17·0483 written to 2 significant figures is

  A 17 B 17·0 C 17·04 D 17·05 E 17·048

9 0·060 794 written to 3 significant figures is

  A 0·060 B 0·061 C 0·0607 D 0·0608 E 0·060 75

10 $\frac{1}{6}$ written as a decimal to 3 significant figures is

    A 0·0167   B 0·0166   C 0·167   D 0·166   E 0·017

11 The estimation of $\dfrac{19\cdot2\times0\cdot432}{4410}$ to 1 significant figure is

    A 2   B 0·2   C 0·02   D 0·002   E 0·0002

12 Given that $\dfrac{52\times2350}{650}=188$ then $\dfrac{520\times2\cdot35}{0\cdot065}$ is exactly

    A 1·88   B 188   C 1880   D 18 800   E 188 000

13 $28\cdot02\div0\cdot0003$ is exactly

    A 9·34   B 93·4   C 934   D 9340   E 93 400

14 $1\cdot6\times0\cdot0005$ is exactly

    A 0·08   B 0·008   C 0·0008   D 0·000 08   E 0·000 008

15 In measuring the length of a square to be 2·5 cm, the greatest possible percentage error is

    A 1%   B 2%   C 4%   D 5%   E 0·1%

16 The length and breadth of a rectangle are given as 4 cm and 2 cm respectively, to the nearest cm. The smallest possible area of the rectangle is, in cm^2,

    A 8   B 5·25   C 11·25   D 4·75   E 6

17 The value of 5^{-2} is

    A $\frac{1}{10}$ B $-\frac{1}{10}$ C -25 D $\frac{1}{25}$ E $-\frac{1}{25}$

18 The value of $27^{\frac{2}{3}}$ is

    A 18   B $\frac{1}{9}$ C 6   D 9   E none of them.

19 The value of $\left(\dfrac{16}{81}\right)^{-\frac{3}{4}}$ is

    A $-\frac{8}{27}$ B $\frac{27}{8}$ C $-\frac{4}{27}$ D $\frac{27}{4}$ E $\frac{4}{9}$

20 The value of $\dfrac{2\times2^0}{2^{-2}}$ is

    A 1   B 2   C 4   D 8   E 16

21 If $a=4$, $b=3$, $x=-1$ and $y=0$, then a^x+b^y is

    A $1\frac{1}{4}$ B $2\frac{1}{4}$ C -2 D $\frac{1}{6}$ E -4

22 $0\cdot000\ 064\ 3$ written in standard form is

    A $64\cdot3\times10^{-6}$ B $0\cdot643\times10^{-4}$ C $6\cdot43\times10^{-4}$
    D $6\cdot43\times10^{-5}$ E $0\cdot643\times10^{-5}$

23 If $a=6\times10^6$ and $b=3\times10^{-3}$, then ab is

    A $1\cdot8\times10^3$ B $1\cdot8\times10^2$ C $1\cdot8\times10^{-8}$ D $1\cdot8\times10^{-4}$
    E $1\cdot8\times10^4$

24 The value of $\dfrac{2\times10^6}{4\times10^{-2}}$ is

    A 5×10^5 B 5×10^7 C 5×10^{-2} D 5×10^{-4}
    E 5×10^{-3}

25 The value of $(2\cdot6\times10^3)+(2\cdot0\times10^2)$ is

    A $4\cdot6\times10^5$ B $4\cdot6\times10^6$ C $2\cdot8\times10^3$ D $2\cdot62\times10^6$
    E none of them.

26 $\dfrac{3}{0\cdot000\ 15}$ is, in standard form,

    A 5×10^4 B 2×10^3 C 5×10^3 D 5×10^{-4} E 2×10^4

27 If $\sqrt{6\cdot5}=2\cdot55$ and $\sqrt{65}=8\cdot06$, each given to 3 significant figures, then $\sqrt{650\ 000}$ is

    A 255 B 806 C $25\cdot5$ D $80\cdot6$ E 2550

28 The best approximation to $\sqrt{0\cdot0005}$ is

    A $0\cdot07$ B $0\cdot7$ C $0\cdot02$ D $0\cdot2$ E $0\cdot03$

29 Which of the following square roots can be deduced from the fact that $\sqrt{46}=6\cdot78$, to 3 significant figures?

    1 $\sqrt{0\cdot046}$ 2 $\sqrt{46\ 000}$ 3 $\sqrt{0\cdot0046}$

    A 1 only B 2 only C 3 only D 1,2 E 2,3

30 The value of $\sqrt[4]{16\times3^8}$ is

    A 36 B 18 C 324 D 24 E 6

31 The best approximation to $\sqrt{(0\cdot4)^2+(0\cdot1)^2}$ is

A $0\cdot41$ B $1\cdot3$ C $0\cdot041$ D $0\cdot13$ E $0\cdot52$

32 The value of $(0\cdot006)^2$ is

A $0\cdot0012$ B $0\cdot0036$ C $0\cdot000\,36$ D $0\cdot000\,12$
E $0\cdot000\,036$

33 If $2\cdot6^2=6\cdot76$ to 3 significant figures then $26\,000^2$ is

A $6\cdot76\times10^4$ B $6\cdot76\times10^6$ C $6\cdot76\times10^8$ D $6\cdot76\times10^{10}$
E $6\cdot76\times10^{12}$

34 If $a=\sqrt{10}$, $b=\sqrt{100}$, $c=\dfrac{1}{\sqrt{0\cdot01}}$, $d=\dfrac{1}{0\cdot01}$, which of the

following is (are) true?

1 $b=c$ 2 $d<a$ 3 $d>c$

A 1 only   B 2,3   C 1,2   D 1,3   E 1,2,3

35 The reciprocal of $0\cdot005$ is

A 2   B 20   C 200   D 2000   E 20 000

36 The reciprocal of $0\cdot0002$ in standard form is

A 2×10^{-4} B 2×10^5 C 5×10^{-4} D 5×10^3
E 5×10^4

37 If $a=\frac{1}{38}$ then

A $0\cdot002<a<0\cdot003$.   B $0\cdot02<a<0\cdot03$.   C $0\cdot2<a<0\cdot3$.
D $2<a<3$.   E $20<a<30$.

38 If $\log_{10}5\cdot6=0\cdot721$, then $\log_{10}0\cdot056$ is

A $0\cdot007\,21$ B $0\cdot0721$ C $2\cdot721$ D $\bar{2}\cdot721$ E $\bar{1}\cdot721$

39 If $\log 9=0\cdot954$ and $\log 2=0\cdot301$, which of the following statements is (are) true?

1 $\log 900=95\cdot4$
2 $\log 18=0\cdot954+0\cdot301$
3 $\log 3=0\cdot477$

A 1 only   B 2 only   C 2 and 3   D 1 and 2   E 1,2 and 3

40 If $\log_{10}2 = 0 \cdot 301$, then $\log_{10}16$ is

A $0 \cdot 301 \times 2$ B $0 \cdot 301 \times 8$ C $0 \cdot 301 \times 4$ D $0 \cdot 301 \times 3$
E none of them.

41 The number of cm in $2 \cdot 4$ km is

A 240   B 2400   C 24 000   D 240 000
E 2 400 000

42 $3 \cdot 4$ m^2 changed into cm^2 is

A $3 \cdot 4 \times 10^2$ B $3 \cdot 4 \times 10^3$ C $3 \cdot 4 \times 10^4$ D $3 \cdot 4 \times 10^5$
E $3 \cdot 4 \times 10^6$

43 3 460 000 cm^3, when expressed as m^3, is

A $0 \cdot 346$ B $3 \cdot 46$ C 346   D 3460
E 34 600

44 If the mass of 1 cm^3 of water is 1 g, then the mass in kg of $\frac{3}{4}$ of a litre of water is

A $0 \cdot 0075$ B $0 \cdot 075$ C $0 \cdot 75$ D $7 \cdot 5$ E 75

45 It is estimated that a 75 cl bottle of wine provides 6 glasses of wine. The number of millilitres in each glass should be

A $12 \cdot 5$ B 125   C 1250   D $105 \cdot 5$ E 1055

46 Given that 1 litre is 1000 cm^3 and that the mass of 1 cm^3 of water is 1 g, the mass of $2 \cdot 4$ litres of water, in kg, is

A $0 \cdot 024$ B $0 \cdot 24$ C $2 \cdot 4$ D 24   E 240

47 A cuboid has a base 10 cm by 4 cm. Its volume is the same as that of a cube of 8 cm. The height of the cuboid, in cm, is

A 14   B $1 \cdot 6$ C $0 \cdot 6$ D 12   E $12 \cdot 8$

48 A rectangle has dimensions 240 mm by 150 mm. The area, in cm^2, is

A 36   B 360   C 3600   D 600   E 6000

49 A rectangle has dimensions 27 cm by 12 cm. The length of the side of the square, in cm, which has the same area, is

A 18   B 12   C 16   D 20   E 24

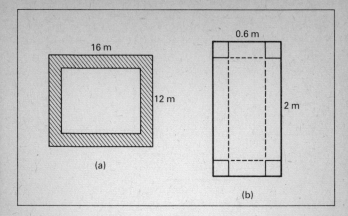

(a)

(b)

Figure 30

50 In figure 30(a), the shaded area between the two rectangles has a constant width of 2 m. The shaded area, in m², is

A 100 B 96 C 90 D 52 E 27

51 Figure 30(b) shows a rectangle 2 m by 0·6 m. From each corner, a square of side 10 cm is cut. The remainder is folded along the dotted lines to make an open rectangular box. Its volume, in m³, is

A 7·2 B 0·72 C 0·072 D 0·0072
E none of them.

52 In figure 31(a), *WXYZ* is a parallelogram with *WX* = 5 cm and *XY* = 13 cm. The area, in cm², of the parallelogram is

A 65 B 60 C 132 D 32·5 E 45

53 In figure 31(b), *ABXY* is a square. *P* and *Q* are the midpoints of *AY* and *XY*. The ratio of the area of triangle *PBQ* to that of the square is

A 1 : 2 B 2 : 3 C 1 : 4 D 3 : 8
E 4 : 9

54 In figure 31(c), the area of the trapezium, in cm², is

A 30 B 100 C 50 D 80·5 E 80

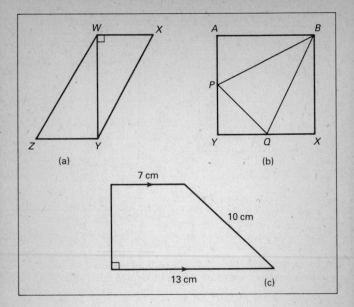

Figure 31

55 A rectangular area, 360 cm by 168 cm, is to be covered with square tiles. No tile is to be cut and all of the tiles are to be the same size. The side of the largest tile that can be used, in cm, is

A 8   B 12   C 16   D 24   E 32

56 Each side of a regular hexagon is 2 cm. The area of the hexagon, in cm², is

A $6\sqrt{3}$ B $12\sqrt{3}$ C 12   D 36   E 24

57 Figure 32 represents a prism. Its volume, in cm³, is

A 48   B 20   C 24   D 40   E 16

58 A pyramid has a rectangular base 10 cm by 6 cm. The volume of the pyramid is 180 cm³. The height, in cm, is

A 3   B 12   C 9   D 6   E 8

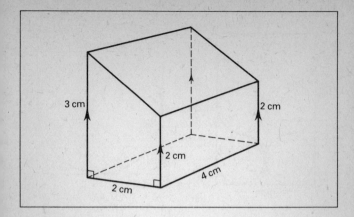

Figure 32

59 The diameter of a circle is d, the radius r and the circumference is C. The value of π is given by

A $\dfrac{C}{d}$ B $\dfrac{2r}{C}$ C $\dfrac{2C}{r}$ D $\dfrac{C}{r}$ E $\dfrac{C}{2d}$

60 The circumference of a circle is 66 cm. Taking $\pi = 3\frac{1}{7}$ the radius of the circle, in cm, is

A 21 B 10·5 C 5·25 D 14 E 7

61 The area of a circle is 616 cm². Taking $\pi = 22/7$, the diameter of the circle, in cm, is

A 42 B 88 C 28 D 21 E 14

62 The circumference of a circle is 12π cm. The area, in cm², is

A 36π B 12π C 100π D 144π
E none of them.

63 In figure 33(a), $ABCD$ is a rectangle. $AB = 42$ cm and $AD = 28$ cm. Six equal circles of radius 7 cm are cut from the rectangle. The area of the remainder, in cm², is   (Take $\pi = 3\frac{1}{7}$)

A 1022 B 714 C 350 D 252 E 420

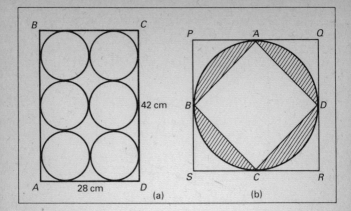

Figure 33

64 In figure 33(b), *PQRS* is a square. The circle with radius 7 cm
    touches each side of the square at *A,B,C* and *D*. The shaded
    area, in cm², is   (Take $\pi = 3\frac{1}{7}$)

    A 126   B 105   C 98   D 56   E 42

65 The total surface area of a solid cylinder with radius 7 cm and
    height 13 cm, in cm², is   (Take $\pi = 22/7$)

    A 660   B 726   C 440   D 594   E 880

66 The volume of the cylinder with radius *x* cm and height *x* cm,
    in cm³, is

    A $3\pi x$ B πx^3 C $2\pi x^2$ D $3\pi x^2$ E none of them.

67 A solid cone has a radius of 14 cm and a height of 6 cm. It is
    made of material of which 1 cm³ has a mass of 2 g. The mass,
    in kg, of the cone is   (Take $\pi = 3\frac{1}{7}$)

    A 7·386   B 73·86   C 2·464   D 24·64   E 0·7386

68 A metal sphere of radius 6 cm is melted down and made into
    small spheres of radius 0·1 cm. The number of spheres made
    is

    A 36 000   B 21 600   C 2160   D 3600
    E 216 000

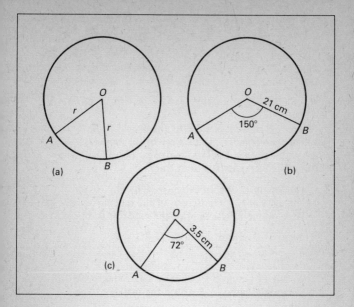

Figure 34

69 In figure 34(a), O is the centre of the circle and chord AB is equal to the radius r. The ratio $\dfrac{\text{minor arc } AB}{\text{major arc } AB}$ is

A $\frac{1}{6}$ B $\frac{1}{8}$ C $\frac{2}{5}$ D $\frac{1}{4}$ E $\frac{1}{5}$

70 In figure 34(b), O is the centre of the circle with radius 21 cm. The area of the sector AOB, in cm^2, is   (Take $\pi = 3\frac{1}{7}$)

A 924 B 1155 C 55 D 577·5 E 110

71 In figure 34(c), O is the centre of the circle. The area of the major sector AOB, in cm^2, is

A 7·7 B 30·8 C 17·6 D 4·4 E 12

72 Arthur, Bill and Cecil share £450 amongst them in the ratio 2 : 3 : 4. Cecil's share is

A £150 B £300 C £200 D £250 E £100

73 If $p : q = 3 : 5$ and $r : q = 1 : 4$ then $p : r$ is

   A 3 : 1   B 5 : 4   C 3 : 4   D 5 : 1   E 12 : 5

74 A sum of money is shared between X and Y in the ratio 5 : 3. If Y's share is £27, then X's share is

   A £45   B £72   C £216   D £54   E £14

75 The model of a car is made to a scale of 1 : 100. The real car is 4·6 metres long. The windscreen of the model is 1·82 cm^2 and the volume of the boot of the real car is 0·32 m^3. Which of the following statements is (are) true?

   1 The length of the model car is 4·6 cm.
   2 The area of the windscreen of the real car is 18·2 m^2.
   3 The volume of the boot of the model is 0·32 cm^3.

   A 1 only   B 2 only   C 1,3   D 2,3
   E 1,2,3

76 The scale of a map is 1 : 25 000. The distance on the ground, in km, represented by 8 cm on the map, is

   A 0·2   B 2   C 20   D 200   E 2000

77 The scale of a map is 1 : 10 000. The area on the map, in cm^2, which represents an area of 2·4 km^2 on the ground, is

   A 0·024   B 0·24   C 2·4   D 24   E 240

78 If $p = 0·035q$, then p expressed as a percentage of q, is

   A 3·5   B 35   C 0·35   D 0·035
   E none of them.

79 When £3.50 is deducted from £25, the percentage remaining is

   A 1·4   B 14   C 30   D 98·6   E 86

80 When £80 is reduced by 70%, it becomes

   A £56   B £5·60   C £24   D £2·40   E £54

81 The price of a coat after being increased by 10% is £55. The pre-increase price is

   A £50   B £49·50   C £54   D £44   E £48·50

82 A car depreciates 15% of its value each year. At the end of a certain year it is worth £3400. What was it worth at the beginning of that year?

A £3910   B £4000   C £3570   D £3579
E none of them.

83 The cost price of an article is £6·50. It is sold for £9·10. The percentage profit is

A 40   B 65   C 160   D 60   E 140

84 X sells an article to Y at a profit of 30%. Y sells it to Z at a profit of 10% on his cost price. The overall % gain on the article is

A 20   B 40   C 43   D 33   E 63

85 On a certain day, 25% of the buses arriving at a depot are late. Of the remainder, 10% are early. What % of the total number of buses arrive on time?

A 35   B 65   C 82·5   D 67·5   E 32·5

Chapter 13
Statistics

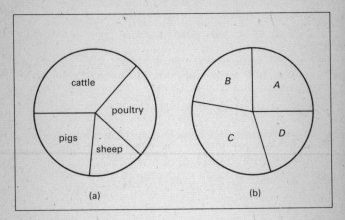

Figure 35

1 The pie chart in figure 35(a) shows the value of the agricultural livestock in a certain country in 1982. 40% of the value is cattle, 25% of the value is poultry, 15% of the value is sheep and the rest is pigs. The size of the angle of the sector representing pigs is

   A 20°   B 36°   C 60°   D 72°   E 90°

2 The results of a survey of 1800 people, asking which of three brands of margarine A, B or C they use, is given in figure 35(b). $\frac{1}{4}$ of the people use brand A; the angle representing those who use brand B is 80°; 570 people use brand C. The number of people who use none of the brands, represented by sector D, is

   A 450   B 380   C 970   D 180   E 540

3 Figure 36(a) shows the result of a survey of the number of people living in each house in a particular road, given as a bar chart. Which of the following statements is (are) true?

1 The total number of houses in the road is 32.
2 The modal number of people per house is 8.
3 The median number of people per house is 3.

A 1 only   B 2 only   C 3 only   D 1 and 3   E 2 and 3

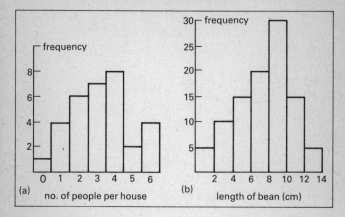

Figure 36

4 The frequency diagram in figure 36(b) shows the length of beans grown under different conditions, experimentally. Which of the following statements is (are) true?
1 The modal class is 8 to 10 cm.
2 The mean length of bean is approximately $7 \cdot 5$ cm.
3 The chart shows that more than half of the beans are over 6 cm long.

A 1 only   B 3 only   C 1 and 2   D 2 and 3   E 1,2 and 3

5 The number of people travelling in each car that passes a checkpoint is given in the frequency table, below.

people per car	1	2	3	4	5	6
frequency	36	24	17	8	4	1

The median number of people per car is

A 1   B 2   C 3   D 4   E 5

6

score	1	2	3	4	5	6
frequency	4	3	3	5	1	4

A die is thrown 20 times and the scores are shown in the table, above. The mean score is

A 1·05   B 3·4   C 11·3   D 4   E none of them.

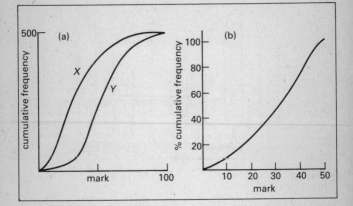

Figure 37

7 Figure 37(a) shows the cumulative frequency curves of the marks obtained by 500 pupils in two tests X and Y. Which of the following statements is (are) true?

 1 The median mark for test Y is lower than that of test X.
 2 If the pass mark is to be the same for both tests, then more pupils will pass test Y than test X.
 3 The same number of pupils are to pass each test. The pass mark on test Y will be higher than that on test X.

 A 1 only   B 2 only   C 3 only   D 1 and 2
 E 2 and 3

8 Figure 37(b) shows the cumulative frequency curve for the marks obtained in an examination. The best estimation for the inter-quartile range is

 A 10   B 15   C 20   D 25   E 30

9 A factory employs 30 men and 20 women. The men's average wage is £110. The women's average wage is £90. The overall average wage is

A £100 B £40 C £102 D £96 E £99

10 The mean of five numbers is 6. When another number is added, the mean becomes 7. The extra number is

A 12 B 6 C 7 D 5 E 14

11 A class contains 10 boys and 20 girls. In a test, their mean score altogether is 7 marks. The mean score of the girls, alone, is 8 marks. The mean score of the boys, alone, is

A 7 B 8 C 6·5 D 5 E 4·5

12 The mean of n numbers is 7. If the number 13 is added, the mean of the $n+1$ numbers is 8. The value of n is

A 9 B 8 C 7 D 6 E 5

13 The mean of x articles is p pence. The mean of another y articles is q pence. The mean of all the articles together is

A $\dfrac{x+y}{p+q}$ B $\dfrac{xp+yq}{x+y}$ C $\dfrac{xq+yp}{x+y}$ D $\dfrac{p+q}{x+y}$ E $\dfrac{xq+yp}{xy}$

14 Two dice are thrown together. Which of the following statements is (are) true?

1 The probability of at least one of them being a 5, is $\frac{2}{36}$.
2 The probability that neither of them is a 4, is $\frac{25}{36}$.
3 The probability that the total score is 6, is $\frac{5}{36}$.

A 1 only B 2 only C 1,2 D 2,3 E 1,2 and 3

15 $N=\{1,2,3,4,5,6,7,8,9,10\}$. An integer is chosen at random from N. The probability that it is any of, an odd number, a multiple of five, or both of these, is

A $\frac{3}{10}$ B $\frac{5}{10}$ C $\frac{6}{10}$ D $\frac{8}{10}$ E $\frac{9}{10}$

16 A number is chosen at random from the numbers 3,3,3,5,5,7,9. The probability that it is the mean of the numbers, is

A $\frac{1}{3}$ B $\frac{2}{5}$ C $\frac{3}{7}$ D $\frac{3}{5}$ E $\frac{2}{7}$

17 5 boys toss a coin in turn. What is the probability that the 4th boy tosses a head?

A $\frac{1}{32}$ B $\frac{1}{16}$ C $\frac{1}{8}$ D $\frac{1}{4}$ E $\frac{1}{2}$

18 The probability that a marksman will hit a target is p. The probability that he misses with his first two shots, is

A p^2 B $(1-p)^2$ C $1-p^2$ D $2p$ E $2-2p$

19 Two cards are drawn from a pack of 52 playing cards without replacing them. The probability that both cards are Queens, is

A $\frac{2}{13}$ B $\left(\frac{1}{13}\right)^2$ C $\left(\frac{1}{4}\right)^2$ D $\frac{1}{13 \times 17}$
E none of them.

20 A bag contains 12 red balls, 5 blue balls and 8 white balls. Two balls are chosen together at random from the bag. The probability that one ball is red and one ball is blue, is

A $\frac{17}{25}$ B $\frac{1}{10}$ C $\frac{1}{5}$ D $\frac{8}{25}$ E $\frac{12}{125}$

21 The probability that A beats B in a game is $\frac{3}{4}$. If they play three games, the probability that B wins the first two games and loses the third, is

A $\frac{9}{10}$ B $\frac{7}{16}$ C $\frac{9}{64}$ D $\frac{1}{16}$ E $\frac{3}{64}$

22 The probability that A is late for school on any day is $\frac{4}{5}$. The probability that B is late on any day is $\frac{5}{7}$. The probability that both of them arrive at school on time on a Friday, is

A $\frac{9}{12}$ B $\frac{20}{35}$ C $\frac{2}{25}$ D $\frac{5}{35}$ E $\frac{2}{35}$

23 An ordinary die is thrown and a coin tossed by two different people at the same time. Which of the following statements is (are) true?

1 The probability of a 6 or a head coming up is $\frac{1}{6} + \frac{1}{2}$.
2 The probability of a 6 and a head coming up is $\frac{1}{12}$.
3 The probability of a head and not a 6 coming up is $\frac{11}{12}$.

A 1 only   B 2 only   C 1,2   D 2,3   E 1,3

Chapter 14
Trigonometry (1)

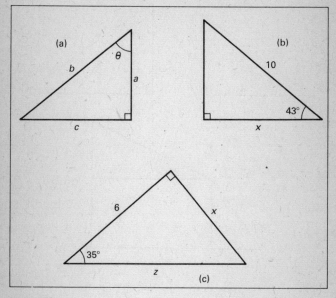

Figure 38

1 In figure 38(a), $\sin \theta$ is

   A $\dfrac{b}{c}$ B $\dfrac{a}{c}$ C $\dfrac{a}{b}$ D $\dfrac{c}{b}$ E $\dfrac{b}{a}$

2 In figure 38(b), given that $\sin 43° = 0\cdot682$, $\cos 43° = 0\cdot731$ and $\tan 43° = 0\cdot933$, the length of x, in cm, is

   A $7\cdot31$ B $6\cdot82$ C $9\cdot33$ D $5\cdot20$ E $68\cdot2$

3 In figure 38(c), x is equal to

   A $6 \tan 55°$ B $6 \sin 35°$ C $6 \cos 35°$ D $6 \tan 35°$
   E none of them.

4 In the same figure, (38c), the value of z is

A $6 \sin 35°$ B $\dfrac{6}{\cos 35°}$ C $\dfrac{6}{\sin 35°}$ D $\dfrac{6}{\tan 35°}$

E $6 \tan 35°$

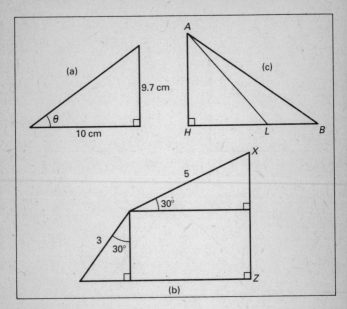

Figure 39

5   $\sin \theta = 0·97,$ $\cos \theta = 0·97,$ $\tan \theta = 0·97,$
    $\theta = 76°.$ $\theta = 14°.$ $\theta = 44°.$

where θ is given to the nearest degree. Using this information, in figure 39(a), the value of θ is

A $76°$ B $14°$ C $44°$ D $46°$ E $24°$

6 In figure 39(b), XZ is

A $3 \sin 30° + 5 \sin 60°$ B $3 \sin 60° + 5 \cos 30°$
C $3 \cos 60° + 5 \sin 30°$ D $3 \cos 30° + 5 \cos 60°$
E none of them.

7 In figure 39(c), if $BL = 8$ cm, $LH = 12$ cm and tan $ABH = \frac{1}{3}$, then tan ALH is

A $\frac{2}{3}$ B $\frac{1}{3}$ C $\frac{5}{12}$ D $\frac{1}{60}$ E none of them.

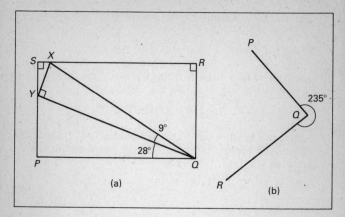

(a)

(b)

Figure 40.

8 In figure 40(a), $XQ = 1$ unit. Which of the following is (are) true?

1 $YQ = \sin 9°$ 2 $PQ = \cos 9° \cos 28°$
3 $SY = \cos 28° \sin 9°$

A 1 only B 2 only C 3 only D 1 and 2
E 2 and 3

9 From a point P, 20 metres from the foot F of a vertical tower, the angle of elevation of the top T is 14°. The height of the tower FT, in metres, is

A 20 tan 14° B 20 tan 76° C 20 cos 14° D 20 sin 14°
E none of them.

10 From the top of a cliff, height h metres, the angle of depression of a boat at sea is θ degrees. The horizontal distance of the boat from the foot of the cliff is

A $h \sin \theta$ B $h \tan \theta$ C $\dfrac{h}{\cos \theta}$ D $\dfrac{h}{\tan \theta}$ E $h \cos \theta$

11 In figure 40(b), the bearing of P from Q is 312°. If reflex angle PQR is 235°, the bearing of R from Q is

A 007°   B 187°   C 083°   D 125°   E 215°

12 The bearing of a point Y from a point X is 130°. The bearing of X from Y is

A 050°   B 040°   C 310°   D 230°   E 220°

13 The bearing of a point L from A is 070°. The bearing of H from A is 150°. If $AL = LH$, the bearing of L from H is

A 080°   B 220°   C 230°   D 130°   E 050°

14 A man in a desert walks 5 km on a bearing 044°, followed by 12 km on a bearing 134°. His distance now from his starting point, in km, is

A 5   B 12   C 7   D 13   E 17

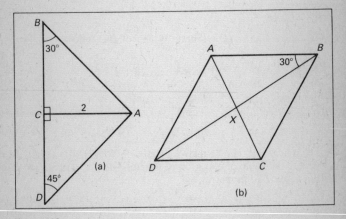

Figure 41

15 In figure 41(a), which of the following statements is (are) true?

1 $BC = 2\sqrt{3}$ cm   2 $AD = 2\sqrt{2}$ cm   3 $AB = \dfrac{2}{\sqrt{3}}$ cm

A 1 only   B 1,2   C 1,3   D 1,2,3   E 3 only

16 In figure 41(b), *ABCD* is a rhombus. Which of the following statements is (are) true, if $AC = 12$ cm?

   1 $XB = 12\sqrt{3}$ cm   2 $AB = 12$ cm   3 $DX = 6\sqrt{3}$ cm

   A 1 only   B 2 only   C 1,3   D 2,3   E 1,2,3

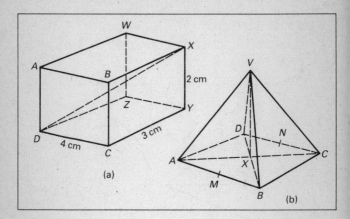

(a)

(b)

Figure 42

17 In figure 42(a), the cuboid has dimensions $4 \times 3 \times 2$ cm. If the angle between the diagonal *XD* and the base *DCYZ* is θ, then tan θ is

   A $\frac{2}{4}$ B $\frac{2}{5}$ C $\frac{2}{3}$ D $\frac{3}{4}$ E $\frac{4}{3}$

18 The right pyramid in figure 42(b) has a square base. The triangular faces are equilateral. *M* and *N* are the midpoints of *AB* and *DC* respectively. Which of the following statements is (are) true?

   1 The inclination of *VA* to the base is given by angle *VAB*.
   2 The inclination of face *VAB* to the base is given by angle *VMX*.
   3 The angle between faces *VAB* and *VDC* is angle *MVN*.

   A 1 only   B 2 only   C 1,2   D 2,3   E 1,2,3

19 In figure 43(a), *XYZ* is a triangular courtyard and *WY* is a vertical pole. Angle $XYZ = 90°$, the angle of elevation of *W*

from Z is 21°, angle $XZY=47°$ and $XZ=20$ m. The height of the pole, in metres, is

A 20 sin 47° cos 21°   B 20 cos 47° sin 21°
C 20 sin 47° tan 21°   D 20 tan 47° sin 21°
E 20 cos 47° tan 21°

20 A ship's position is latitude 50°N, longitude 20°E. If the radius of the earth is R, the radius of the parallel of latitude on which the ship lies is

A R sin 20°   B R cos 20°   C R sin 50°   D R cos 50°
E R tan 50°

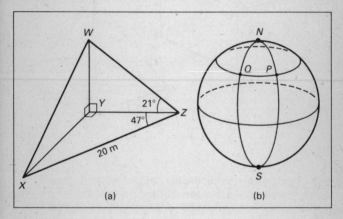

(a)   (b)

Figure 43

21 In figure 43(b), P is at 60°N 20°E and Q is at 60°N 100°W. Which of the following statements is (are) true?
    1 The shortest route from P to Q passes over the North Pole.
    2 The shortest route from P to Q is due west.
    3 The shortest route from P to Q is 3600 nautical miles.

    A 1,2   B 2,3   C 1,3   D 1,2,3   E none of them

22 If the radius of the earth is 6370 km and $\pi=3\frac{1}{7}$, then the distance along the great circle between the points (35°N 113°E) and (55°S 113°E), in km, is

    A 1592·5   B 10 010   C 20 020   D 3185
    E none of them.

82

Chapter 15
Trigonometry (2)

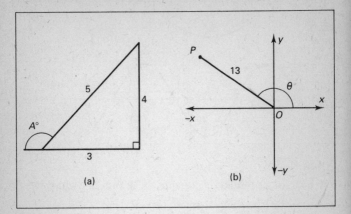

Figure 44

1 In figure 44(a), the value of cos $A°$ is

  A 0·6 B −0·8 C −1·25 D −0·6 E 0·75

2 In figure 44(b), if sin $\theta = \frac{5}{13}$ and $OP = 13$ units, then the co-ordinates of P are

  A (−5,12) B (−12,5) C (5,−12) D (−12,−5)
  E (−5,−12)

3 If sin 30° = $\frac{1}{2}$, then sin 150° is

  A $\frac{1}{2}$ B $-\frac{1}{2}$ C $\frac{\sqrt{3}}{2}$ D $-\frac{\sqrt{3}}{2}$ E 1

4 If cos $\theta = \frac{3}{5}$, then cos (360° − θ) is

  A $\frac{3}{5}$ B $\frac{4}{5}$ C $-\frac{4}{5}$ D $-\frac{5}{3}$ E $-\frac{3}{5}$

5 Which of the following statements is (are) true?
  1 sin 60° = sin 120°   2 tan 30° = tan 150°
  3 cos 143° = cos 217°

  A 1 only B 2 only C 3 only D 1 and 2
  E 1 and 3

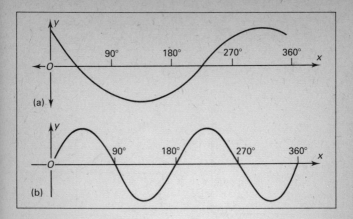

Figure 45

6 Which of the following graphs is drawn in figure 45(a)?

  A $\sin x°$ B $\cos x°$ C $\sin x° + \cos x°$
  D $\sin x° - \cos x°$ E $\cos x° - \sin x°$

7 The graph in figures 45(b) is

  A $\sin 2x°$ B $\cos 2x°$ C $2 \sin x°$ D $2 \cos x°$
  E $2(\sin x° - \cos x°)$

8 Given that $\sin 30° = \dfrac{1}{2}$, $\sin 60° = \dfrac{\sqrt{3}}{2}$ then $\sin 240°$ is

  A $\dfrac{1}{2}$ B $-\dfrac{1}{2}$ C $\dfrac{\sqrt{3}}{2}$ D $-\dfrac{\sqrt{3}}{2}$ E 1

9 The value of $\cos 30° + \sin 120°$ is

  A 0 B $\sqrt{3}$ C 1 D $\dfrac{\sqrt{3}+1}{2}$ E $\dfrac{\sqrt{3}-1}{2}$

10 If $\tan \theta = \frac{3}{5}$, then $\cos^2 \theta$ is

  A $\frac{25}{16}$ B $\frac{9}{34}$ C $\frac{9}{16}$ D $\frac{25}{34}$ E $\frac{3}{4}$

11 If $\cos^2 x = \frac{1}{5}$ and x is an acute angle, the value of $\sin x°$ is

  A $\sqrt{\dfrac{24}{25}}$ B $\dfrac{1}{\sqrt{5}}$ C $\dfrac{1}{5}$ D $\dfrac{2}{\sqrt{5}}$ E none of them.

12 If $\sin x = \dfrac{1}{\sqrt{3}}$ and x is an obtuse angle, then $\tan x$ is

A $\dfrac{1}{\sqrt{2}}$ B $\sqrt{\dfrac{2}{3}}$ C $-\sqrt{\dfrac{2}{3}}$ D $-\dfrac{2}{3}$ E $\dfrac{-1}{\sqrt{2}}$

13 If $\sin \alpha = \frac{3}{5}$, then $\tan (90 - \alpha)$ is

A $\frac{4}{3}$ B $\frac{3}{4}$ C $\frac{4}{5}$ D $\frac{5}{4}$ E $\frac{5}{3}$

14 If $\cos 60° = \frac{1}{2}$, then $\tan 330°$ is

A 2 B $-\sqrt{3}$ C $\dfrac{1}{\sqrt{3}}$ D $-\dfrac{1}{\sqrt{3}}$ E $\sqrt{3}$

15 If $\cos x = -\dfrac{\sqrt{3}}{2}$ and $0° \leqslant x \leqslant 360°$, then x is

A 150° only B 330° only C 150° and 210°
D 150° and 330° E 30° and 330°

16 If $\tan x = -1$ and $0° \leqslant x \leqslant 360°$, then x is

A 45° only B 135° only C 135° and 225°
D 135° and 315° E none of them.

17 In a triangle ABC, $BC = 10$ cm, $AB = 6$ cm and $AC = 7$ cm. The value of $\cos BAC$ is given by

A $\dfrac{6^2 + 7^2 - 10^2}{2 \times 6 \times 7}$ B $\dfrac{6^2 + 10^2 - 7^2}{2 \times 6 \times 10}$

C $\dfrac{7^2 + 10^2 - 6^2}{2 \times 7 \times 6}$ D $\dfrac{6^2 - 7^2 - 10^2}{2 \times 6 \times 7}$

E $\dfrac{6^2 - 10^2 - 7^2}{2 \times 6 \times 10}$

18 In a triangle PQR, angle $PQR = 40°$, angle $PRQ = 20°$ and $PR = 5$ cm. The length of QR is given by

A $\dfrac{5 \sin 40°}{\sin 20°}$ B $\dfrac{5 \sin 120°}{\sin 40°}$ C $\dfrac{5 \sin 40°}{\sin 120°}$

D $\dfrac{5 \sin 20°}{\sin 120°}$ E $\dfrac{5 \sin 20°}{\sin 40°}$

Chapter 16
Calculus

1 If $y = 4x^3 - 2x + 1$, then $\dfrac{dy}{dx}$ is

A $12x - 2$ B $12x - 1$ C $12x^2 - 2$ D $12x^2 - 1$
E $4x^2 - 2$

2 The gradient at the point where $x = 5$, on the curve $y = 2x^2 - 3x$, is

A $\frac{1}{17}$ B 17   C 7   D 5   E $\frac{1}{7}$

3 If $y = \dfrac{1}{x^3}$, then $\dfrac{dy}{dx}$ is

A $\dfrac{1}{3x^2}$ B $-\dfrac{3}{x^2}$ C $-\dfrac{1}{3x^2}$ D $-\dfrac{1}{3x^4}$ E $-\dfrac{3}{x^4}$

4 The gradient at the point $(-3, -\frac{2}{3})$ on the curve $y = \dfrac{2}{x}$, is

A $-\frac{2}{9}$ B -2 C $-\frac{1}{18}$ D 2   E $-\frac{9}{2}$

5 The graph $y = x^2 - 3x + 2$ has a turning point where x is

A 1 and 2   B -1 and -2 C $1\frac{1}{2}$ only   D $-1\frac{1}{2}$ only
E 1 only

6 The co-ordinates of the maximum point on the curve $y = x^3 - 3x^2 - 9x$ are

A $(3, -27)$ B $(-1, 5)$ C $(-3, 27)$ D $(1, 9)$
E $(-1, 11)$

7 The co-ordinates of the point on the curve $y = x^2 + 2x$, at which the gradient is -2, are

A $(0, 2)$ B $(-2, 0)$ C $(0, -2)$ D $(\frac{1}{2}, \frac{5}{4})$ E $(-\frac{1}{2}, -\frac{3}{4})$

8 The co-ordinates of the point on the curve $y = 7x - 3x^2$, at which the tangent makes an angle of $45°$ with the positive x-axis, are

A $(0, 5)$ B $(-1, -10)$ C $(\frac{7}{3}, 0)$ D $(1, 4)$
E none of them.

9 If $y = x^3 - x^2$, which of the following statements is (are) true?

1 The curve has a turning point at the origin.
2 When $x = 1$ the gradient is tan 45°
3 When $x = -\frac{1}{2}$ the gradient is positive.

A 1 only   B 2 only   C 1,2   D 2,3   E 1,2,3

10 The distance s metres travelled by a moving point in t seconds is given by $s = 5t^2 + 14t - 6$. The speed, in metres per second, when $t = 0$ is

A −6   B 10   C 24   D 14   E 13

11 The velocity in m/s after t seconds of a point moving in a straight line, is given by $v = 2t - 3t^2$. The acceleration, in m/s^2, when $t = 1$ is

A −2   B −4   C 2   D 4   E −1

12 The distance s metres travelled by a moving point in t seconds is given by $s = t^2 - 8t$. The point is momentarily at rest when $t =$

A 8   B 0   C 4   D 6   E $3\frac{1}{2}$

13 The gradient at any point on a curve is $4x^3 + 2x$. The equation of the curve, where C is a constant, is

A $4x^4 + x^2 + C$ B $x^4 + 2x^2 + C$ C $4x^4 + 2x^2 + C$
D $x^4 + x^2 + C$ E none of them.

14 A graph which passes through the origin has a gradient given by $\dfrac{dy}{dx} = 4x$. The equation of the graph is

A $y = x + 4$ B $x = 2y^2$ C $y = 2x^2$ D $y = 4x^2$
E $x = 4y^2$

15 The gradient at any point on a curve is given by $6x - 5$. The curve passes through the point (2,1). The equation of the curve is

A $y = 6x^2 - 5x - 13$
B $y = 3x^2 - 5x - 1$
C $y = 3x^2 - 5x - 3$
D $y = 6x - 11$
E $y = 6x^2 - 5x - 15$

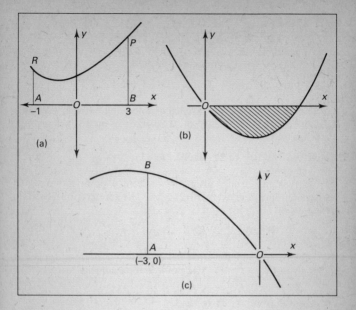

Figure 46

16 In figure 46(a), RP is an arc of the curve $y = x^2 + x + 1$. The area enclosed by $ABPR$ is given by

A $\displaystyle\int_3^{-1} x\,dy$ B $\displaystyle\int_3^{-1} y\,dx$ C $\displaystyle\int_{-1}^3 y\,dx$ D $\displaystyle\int_3^{-1} y^2\,dx$

E $\displaystyle\int_{-1}^3 x^2\,dy$

17 In figure 46(b), the graph of $y = x^2 - 2x$ is given. The shaded area, in square units, is

A -2 B $-\frac{16}{3}$ C $-\frac{7}{3}$ D $-\frac{2}{3}$ E $-\frac{4}{3}$

18 In figure 46(c), the curve is $y = -8x - x^2$. The point A has co-ordinates $(-3, 0)$. The area enclosed by the lines OA, AB and the curve, in square units, is

A 2 B 33 C 15 D 27 E none of them.

19 The value of $\int_{-1}^{2} x^3 \, dx$ is

A $\frac{15}{4}$ B $\frac{17}{4}$ C $-\frac{15}{4}$ D $-\frac{17}{4}$ E none of them.

20 The value of $\int_{1}^{3} \frac{1}{x^2} \, dx$ is

A $-\frac{2}{3}$ B $\frac{26}{27}$ C $\frac{28}{27}$ D $\frac{4}{3}$ E $\frac{2}{3}$

21 A particle P starts from a point O and moves along a straight line so that the velocity, in metres per second, after t seconds, is given by $v = 3t^2 + 2t$. The distance from O, in metres, after 2 seconds is

A 14   B 12   C 10   D 32   E 16 .

22 A particle starts from rest with an acceleration in metres per second per second, after t seconds, given by $a = 3t - 2$. The velocity after 3 seconds is

A 3   B 7   C 7·5   D 11·5   E 1·5

23 A body starts from a fixed point O and moves along a straight line with a velocity after t seconds of $v = (12t - 3t^2)$ metres per second. Which of the following statements is (are) true?

1 The body is momentarily at rest when $t = 12$ seconds.
2 The body is retarding when $t = 3$ seconds.
3 The body travels 16 metres in the 2nd second.

A 1 only   B 2 only   C 1,2   D 1,2,3   E none of them.

24 A particle starts from a fixed point O. It moves in a straight line so that its velocity after t seconds is given by $v = t(8 - t)$ metres per second. The distance travelled in the 3rd second, in metres, is

A $\frac{41}{3}$ B 27   C $\frac{40}{3}$ D 2   E $\frac{121}{3}$

Answers

Sets

1 B, 2 C, 3 A, 4 B, 5 C, 6 A, 7 D, 8 C, 9 E, 10 E, 11 A, 12 A, 13 C, 14 D, 15 C, 16 E, 17 E, 18 D, 19 C.

Basic Algebra

1 E, 2 B, 3 A, 4 C, 5 B, 6 A, 7 D, 8 E, 9 B, 10 C, 11 B, 12 B, 13 B, 14 B, 15 C, 16 B, 17 C, 18 B, 19 A, 20 D, 21 B, 22 E, 23 A, 24 D, 25 A, 26 D, 27 A, 28 B, 29 A, 30 C, 31 E, 32 B, 33 A, 34 A, 35 D, 36 A, 37 B, 38 B, 39 D, 40 E, 41 B, 42 B, 43 B, 44 D, 45 C, 46 A, 47 D, 48 C, 49 E, 50 C, 51 D, 52 C, 53 B, 54 C.

Number Systems

1 C, 2 B, 3 B, 4 D, 5 D, 6 A, 7 D, 8 C, 9 B, 10 A, 11 E, 12 C, 13 C, 14 E, 15 C, 16 D, 17 B, 18 C, 19 E, 20 E, 21 E, 22 A, 23 E, 24 B, 25 B, 26 C, 27 A, 28 C.

Functions (1)

1 D, 2 B, 3 D, 4 B, 5 C, 6 B, 7 C, 8 A, 9 D, 10 E, 11 B, 12 A, 13 B, 14 D, 15 C, 16 E, 17 A, 18 D, 19 B, 20 A, 21 D, 22 C, 23 D, 24 C, 25 E, 26 C, 27 A, 28 B, 29 B, 30 D, 31 E, 32 D.

Functions (2)

1 B, 2 A, 3 D, 4 A, 5 C, 6 B, 7 C, 8 B, 9 A, 10 B, 11 C, 12 E, 13 C, 14 C, 15 E, 16 B, 17 C, 18 D, 19 D, 20 E, 21 B.

Inequalities

1 B, 2 D, 3 D, 4 E, 5 B, 6 D, 7 B, 8 A, 9 A, 10 D, 11 E, 12 A, 13 D, 14 D, 15 B, 16 C.

Matrices

1 D, 2 B, 3 E, 4 A, 5 B, 6 D, 7 C, 8 D, 9 B, 10 C, 11 D, 12 C, 13 E, 14 B, 15 D, 16 B, 17 C, 18 A, 19 D, 20 E.

Elementary Geometry

1 A, 2 B, 3 C, 4 C, 5 D, 6 C, 7 A, 8 D, 9 B, 10 D, 11 B, 12 A, 13 D, 14 C, 15 D, 16 A, 17 E, 18 C, 19 B, 20 E, 21 B, 22 A, 23 C, 24 D, 25 A, 26 E, 27 B, 28 A, 29 D, 30 B, 31 B, 32 D, 33 D.

Transformation Geometry

1 A, 2 D, 3 B, 4 B, 5 A, 6 B, 7 E, 8 D, 9 B, 10 A, 11 B, 12 D, 13 E, 14 E, 15 D, 16 C, 17 E.

Transformation Matrices

1 D, 2 B, 3 A, 4 D, 5 C, 6 D, 7 C, 8 D, 9 E, 10 C, 11 B, 12 D, 13 A.

Vectors

1 A, 2 D, 3 C, 4 D, 5 E, 6 C, 7 E, 8 D, 9 D, 10 C, 11 E, 12 D, 13 A, 14 B, 15 D, 16 B, 17 D, 18 E, 19 A, 20 C, 21 E, 22 B, 23 B.

Basic Arithmetic

1 B, 2 B, 3 E, 4 B, 5 D, 6 B, 7 D, 8 A, 9 D, 10 C, 11 D, 12 D, 13 E, 14 C, 15 B, 16 B, 17 D, 18 D, 19 B, 20 D, 21 A, 22 D, 23 E, 24 B, 25 C, 26 E, 27 B, 28 C, 29 C, 30 B, 31 A, 32 E, 33 C, 34 D, 35 C, 36 D, 37 B, 38 D, 39 C, 40 C, 41 D, 42 C, 43 B, 44 C, 45 B, 46 C, 47 E, 48 B, 49 A, 50 B, 51 C, 52 B, 53 D, 54 E, 55 D, 56 A, 57 B, 58 C, 59 A, 60 B, 61 C, 62 A, 63 D, 64 D, 65 E, 66 B, 67 C, 68 E, 69 E, 70 D, 71 B, 72 C, 73 E, 74 A, 75 C, 76 B, 77 E, 78 A, 79 E, 80 C, 81 A, 82 B, 83 A, 84 C, 85 D.

Statistics

1 D, 2 B, 3 D, 4 E, 5 B, 6 B, 7 E, 8 C, 9 C, 10 A, 11 D, 12 E, 13 B, 14 D, 15 C, 16 E, 17 E, 18 B, 19 D, 20 C, 21 E, 22 E, 23 C.

Trigonometry (1)

1 D, 2 A, 3 D, 4 B, 5 C, 6 D, 7 B, 8 E, 9 A, 10 D, 11 B, 12 C, 13 E, 14 D, 15 B, 16 D, 17 B, 18 D, 19 E, 20 D, 21 B, 22 B.

Trigonometry (2)

1 D, 2 B, 3 A, 4 A, 5 E, 6 E, 7 A, 8 D, 9 B, 10 D, 11 D, 12 E, 13 A, 14 D, 15 C, 16 D, 17 A, 18 B.

Calculus

1 C, 2 B, 3 E, 4 A, 5 C, 6 B, 7 B, 8 D, 9 E, 10 D, 11 B, 12 C, 13 D, 14 C, 15 B, 16 C, 17 E, 18 D, 19 A, 20 E, 21 B, 22 C, 23 B, 24 A.

Solutions

There now follow some hints and solutions to some of the questions.

Sets

5 The shaded area can be described as, "in set P and not in Q and not in R". In symbols, this is $P \cap Q' \cap R'$.

18 Statement 1 $n(C \cap T) = 3 + 8 = 11$. $\therefore$ 1 is true.

Statement 2 $n(C \cup P \cup T) = 1 + 3 + 4 + 7 + 8 + 5 + 6 = 34$
$n(C \cup P \cup T)' = 40 - 34 = 6$. $\therefore$ 2 is true.

Statement 3 $n(C) = 3 + 4 + 7 + 8 = 22$. $\therefore$ 3 is false.

Algebra

19 Factorizing $6x^2 - 11x - 10$ gives $(3x + 2)(2x - 5)$. Factors of a number divide exactly into the number $\therefore$ $(3x + 2)$ is the only statement which is true.

40 Multiplying $(3x - 2)(x + 2)$ gives $3x^2 + 4x - 4$, which is the right hand side. The left hand side is exactly the same as the right hand side, i.e. it is true for all values of x.

Number Systems

5 The set of integers $= \{ \ldots -3, -2, -1, 0, 1, 2, 3 \ldots \}$.

Now, $\dfrac{n+1}{2}$ is an integer $\Rightarrow$ 2 divides exactly into $n + 1 \Rightarrow n + 1$ is even $\Rightarrow n$ is odd. $\therefore$ D is the required suggestion.

25 In the table, column b and row b have the set $\{a, b, c, d\}$ in that order. This makes b the identity element.

Functions (1)

2 Remember that the domain $\rightarrow$ the range, or the domain contains the set of x values and the range the set of y values.

Here $y = x^2$ or $x = \pm\sqrt{y}$. Substitute each value of y in turn. When $y = 0$ $x = 0$; $y = 1$, $x = \pm 1$; $y = 2$, $x = \pm\sqrt{2}$; $y = 3$, $x = \pm\sqrt{3}$; and $y = 4$, $x = \pm 2$. The domain varies between -2 and $+2$ inclusive, making B the correct suggestion.

25 Find the points where $3x+4y=2$ cuts the axes. It cuts the x-axis where $y=0 \Rightarrow 3x=2 \Rightarrow x=+\frac{2}{3}$.
It cuts the y-axis where $x=0 \Rightarrow 4y=2 \Rightarrow y=+\frac{1}{2}$.

Both of the intercepts are on the $+$ axes. Only diagram E has this situation.

Functions (2)

9 The line l cuts the curve at points whose x values are -2 and $+5$. $\therefore$ $x=-2$ and $x=+5$ are the solutions of the required equation. This means that $(x+2)(x-5)$ are the brackets. Multiply them out to find that $x^2-3x-10=0$ is the equation.

19 The gradient on a distance/time graph measures the speed. The gradient is constant, therefore the speed is constant, making A and B false.

The distance travelled $= \dfrac{\text{speed}}{\text{time}} = \dfrac{200}{16-8} = \dfrac{200}{8} = 25$ m/s.

This makes C false and D true.

Inequalities

4 $3x+8 < 3(x+4) \Rightarrow 3x+8 < 3x+12 \Rightarrow 8 < 12$. This is true for any value of x.

16 The shaded area is completely defined by $x+y > 3$; $y > x$; $y < 3$. The only **pair** of inequalities to be listed is in Statement 3.

Matrices

3 $B^2 = B \times B = \begin{pmatrix} 3 & 4 \\ -1 & 2 \end{pmatrix} \begin{pmatrix} 3 & 4 \\ -1 & 2 \end{pmatrix} = \begin{pmatrix} 5 & 20 \\ -5 & 0 \end{pmatrix}$

B^2 means multiply the matrix B by itself. Do not just square each element of B.

10 The determinant of the matrix is $(7 \times 2) - (5 \times 3) = -1$.

Hence the inverse is $-1 \begin{pmatrix} 2 & -5 \\ -3 & 7 \end{pmatrix} = \begin{pmatrix} -2 & 5 \\ 3 & -7 \end{pmatrix}$.

C is correct. Take care of the minus value of the determinant.

Elementary Geometry

7 The triangles in diagrams 1 and 4 have 2 sides and the included angle equal, $\therefore$ the triangles are congruent. The triangles in diagrams 2 and 3 have three equal angles, but the equal sides are not in corresponding positions, $\therefore$ the triangles do not pass the test of congruence (in fact they are similar).

28 angle $QPS = 90°$ (the angle in a semicircle is a right angle).
angle $QPR = 90° - 48° = 42°$.
angle $x = 42°$ (angles QPR and x stand on the same arc QR).

Transformation Geometry

11 The enlarged parallelogram is similar to the original. Similar figures with sides in the ratio $a : b$ have areas in the ratio $a^2 : b^2$. Here the sides are in the ratio 1 : 4, so the areas are in the ratio 1 : 16. The required area is $9 \times 16 = 144$ cm^2.

Transformation Matrices

11 The transformation UV means V followed by U.

Hence, $UV = \begin{pmatrix} 2 \\ -1 \end{pmatrix} \rightarrow \begin{pmatrix} 2 \\ -1 \end{pmatrix} + \begin{pmatrix} 1 \\ 2 \end{pmatrix} = \begin{pmatrix} 3 \\ 1 \end{pmatrix}$

then, $\begin{pmatrix} 3 \\ 1 \end{pmatrix} \rightarrow \begin{pmatrix} 1 & 2 \\ 0 & 1 \end{pmatrix}\begin{pmatrix} 3 \\ 1 \end{pmatrix} = \begin{pmatrix} 5 \\ 1 \end{pmatrix}$

The image is (5,1).

Vectors

4 Statement 1 $\overrightarrow{OX} = \overrightarrow{OB} + \overrightarrow{BX} = \mathbf{a} + \mathbf{b} \therefore 1$ is true.

Statement 2 Using vector addition, $\overrightarrow{OA} + \overrightarrow{AB} = \overrightarrow{OB}$
$\overrightarrow{AB} = \overrightarrow{OB} - \overrightarrow{OA} = \mathbf{b} - \mathbf{a}. \therefore 2$ is false.

Statement 3 $|\mathbf{a}| = |\mathbf{b}| \Rightarrow OA = OB$, making $OAXB$ a rhombus. The diagonals of a rhombus bisect at right angles, $\therefore \overrightarrow{AB}$ and $\overrightarrow{OX}$ are perpendicular vectors, $\therefore 3$ is true.

Arithmetic

6 Let $n = $ number of eligible voters, $\therefore \frac{5}{8}n$ actually vote.

A and B receive $\frac{2}{5} + \frac{1}{3} = \frac{6+5}{15} = \frac{11}{15}$ of the actual vote.

$\therefore C$ receives $\frac{4}{15}$ of the actual vote, i.e. $\frac{4}{15} \times \frac{5}{8}n = \frac{1}{6}n$

$\therefore C$ receives $\frac{1}{6}$ of the eligible vote.

Statistics

18 The probability of a hit is p, $\therefore$ the probability of a miss is $1 - p$. Each shot is independent of the next, $\therefore$ the probability of two misses is $(1-p)(1-p)$ or $(1-p)^2$.

Trigonometry (1)

7 $\dfrac{AH}{HB}= \tan ABH$. $\dfrac{AH}{HB}=\dfrac{1}{5} \Rightarrow AH=\dfrac{20}{5}=4$ cm.

In triangle ALH, $\tan ALH=\dfrac{AH}{HL}=\dfrac{4}{12}=\dfrac{1}{3}$.

Trigonometry (2)

16 If $\tan x= -1$, then x is an angle in the 2nd and 4th quadrant. The angle whose tangent is 1, is 45°. The solutions are, $x= 180° - 45°$ and $x= 360° - 45°$, or $x= 135°$ and $315°$.

Calculus

17 The curve cuts the x-axis where $y= 0 \Rightarrow x^2 - 2x= 0 \Rightarrow x(x- 2) = 0 \Rightarrow x= 0$ and $x= 2$, which are the limits for the area.

$$A = \int_0^2 (x^2 - 2x)\ \mathrm{d}x= \left[\frac{x^3}{3} - x^2\right]_0^2$$

$(\frac{8}{3} - 4) - 0= -\frac{4}{3}$. (The negative area denotes that it is below the x-axis.)

Key Facts Educational Aids

KEY FACTS CARDS

Additional Mathematics
Algebra
Arithmetic & Trigonometry
Biology
Chemistry
Computer Studies
Economics
Elementary Mathematics
English Comprehension
English Language
French
General Science

Geography
Geography British Isles
Geometry
German
History (1815-1914)
History (1914-1951)
Human Biology
Modern Mathematics
New Testament
Physics
Technical Drawing

KEY FACTS COURSE COMPANIONS

Additional Mathematics
Algebra
Arithmetic & Trigonometry
Biology
Chemistry
Economics

English
French
Geography
Geometry
Modern Mathematics
Physics

KEY FACTS A-LEVEL BOOKS

Biology
Chemistry

Physics
Pure Mathematics

KEY FACTS O-LEVEL PASSBOOKS

Biology
Chemistry
Computer Studies
Economics
English Language
French
Geography

Geography British Isles
History (Political & Constitutional)
History (Social & Economic)
Human Biology
Modern Mathematics
Physics
Technical Drawing

KEY FACTS A-LEVEL PASSBOOKS

Applied Mathematics
Biology
Chemistry
Economics

Geography
Physics
Pure Mathematics
Pure & Applied Mathematics

KEY FACTS O-LEVEL MODEL ANSWERS

Biology
Chemistry
English Language
French

Geography
History (Social & Economic)
Modern Mathematics
Physics

KEY FACTS REFERENCE LIBRARY

Biology
Chemistry
Geography

History (1815-1914)
Physics
Traditional Mathematics

KEY FACTS O-LEVEL MULTIPLE CHOICE

Biology
Chemistry
Economics
English
French
Geography

Geography British Isles
History (Social & Economic)
Human Biology
Modern Mathematics
Physics

KEY FACTS A-LEVEL WORKED EXAMPLES

Applied Mathematics
Biology
Chemistry
Economics

Geography
Physics
Pure Mathematics
Pure & Applied Mathematics

KEY FACTS DICTIONARIES

Biology
Chemistry
Economics

Geography
Mathematics
Physics